Science in Geography

1

Developments in
geographical method

General editor:
Brian P. FitzGerald

Science in Geography
1
Developments in geographical method

Brian P. FitzGerald

J. J. Aschmann

Oxford University Press 1974

Oxford University Press, Ely House, London W1

Glasgow New York Toronto Melbourne Wellington
Cape Town Ibadan Nairobi Dar es Salaam Lusaka Addis Ababa
Delhi Bombay Calcutta Madras Karachi Lahore Dacca
Kuala Lumpur Singapore Hong Kong Tokyo

PRINTED IN GREAT BRITAIN BY OFFSET LITHOGRAPHY BY
BILLING AND SONS LTD., GUILDFORD AND LONDON

Preface

Geography in schools is at present going through a period of change, a change which represents to many a much-needed overhaul, to others an unnecessary dabble in apparently obscure and complicated statistical techniques. Society today is making demands on education which schools and colleges must recognize by taking part in certain changes, if their students are to become adults equipped to play their full part in the society of tomorrow. Much that is still taught is of questionable relevance to the student's needs.

Geography is a discipline which has been slow to carry through at school level the changes that have been taking place in universities. Because of this, ill-defined subject groups, such as environmental studies, social studies, or interdisciplinary studies, are tending to supersede it in schools. Greater co-operation between subjects is admittedly necessary, but one must be aware of the possibility of geography as a school discipline disappearing completely. This would be highly undesirable from an educational point of view, but quite deserved while geography continues to provide little intellectual rigour. The title of a subject, a mode of inquiry, or a field of knowledge is perhaps unimportant, but we do need an intellectually stronger core to what we teach and learn if the essence of geography is not to vanish from schools and colleges.

The changes that *are* beginning to be introduced are making geography more relevant to the needs of students as they become more involved in urban studies and planning, as they begin to analyse the problems of the developing countries, and as they begin to appreciate the problems of resource conservation. These changes are being accepted and becoming established in schools, but there is still very little said about the nature and philosophy of geography on which these changes depend. It is strongly felt that the sixth-form and college student should understand the more important arguments in this field. Only by involving the student in these issues can we justify what is being studied at Advanced Level and beyond.

The literary, descriptive approach to geography, where geography is treated as an arts subject, still has a definite role to play, but a clearer understanding of the nature of geography is achieved with the scientific approach. Such an approach requires the study of spatial patterns and overall systems of operation; it requires a greater degree of precision in measurement and description; it requires some estimate of the significance of inferences and conclusions drawn from the relationships being studied; and above all it requires an attempt to set up generalized theory from which predictions can be made.

The important tests of the success of the approach are:

(1) whether the student has a better understanding of the organization of society in a spatial (geographical) sense; and

(2) whether he has therefore developed a greater ability to make reasoned decisions based on his improved understanding.

On the first point, generalized 'models' or structures of the working of reality (which form the basis of scientific geography) aid understanding and act as pegs upon which to hang further ideas, concepts, and factual material. As far as the second point is concerned, a scientific approach to geography increases the ability to act upon evidence, and, through the development of general theory, allows decisions to be made which are based on a better understanding of reality. Thus courses of action can be better planned, and a more worthwhile contribution to society can be made.

The four books in the Science in Geography series are:

Developments in Geographical Method by Brian P. FitzGerald
Data Collection by Richard Daugherty
Data Description and Presentation by Peter Davis
Data Use and Interpretation by Patrick McCullagh.

The plan for the series came from an idea of Peter Bryan, from Cambridge-shire High School for Boys, whose advice during all the various stages of producing the books has been of great assistance.

Stonyhurst, August 1973   Brian P. FitzGerald

Contents

             Morphological transformations   44
             Density transformations   53

Chapter 4   **Perception and decision-making in geography**   65
             The perceived environment   65
                The perception of hazards   66
                Migration decisions and the perceived
                environment   70
             Simulation and decision-making   72
             Summary   78

             General Bibliography   81

             Contents of *Science in Geography*,   83
             Books 2, 3, and 4

Chapter 1

The science of geography

Many changes have taken place in geography in recent years. Most of these have so far been at university level, where new courses are appearing in such subjects as 'locational analysis' and 'theoretical urban geography'. New courses include studies of the world's major problems, world resource development and conservation, misuse and pollution of the environment, the development of the 'third' world, and so on.

Courses such as those in theoretical geography and locational analysis reflect a more fundamental change in the basic philosophy and methodology of the subject than do the changes in emphasis from, say, regional geography to the study of world problems. It is with the former changes that this book, and the other three titles in the *Science in Geography* series, are concerned.

Some of the basic changes in methodology and thinking are also appearing at school level, but at whatever level they appear they tend to cause some concern. This is largely brought about by the feeling that 'modern geography' is somehow concerned with abstruse statistical techniques, and what is more, statistical techniques for their own sake. More often than not, so the criticism runs, statistical sledge-hammers are being used to crack geographical nuts. Undoubtedly there is some truth behind these criticisms; many learned articles have leaned heavily upon statistical and other mathematical formulae, and both at university and school level statistical courses that appear to have no relevance to geography have been introduced.

This use of abstruse techniques is one that has caused considerable obscuring of the very real and important developments taking place. Unfortunately many geographers are not trained in mathematics, so that the little mathematics that has to be used may cause complete rejection. Many of these changes may appear to be unnecessary efforts to push geography forward as a pseudo-science, dealing with the unreal and theoretical, and ignoring the literary approach which traditionally projects the true charac-

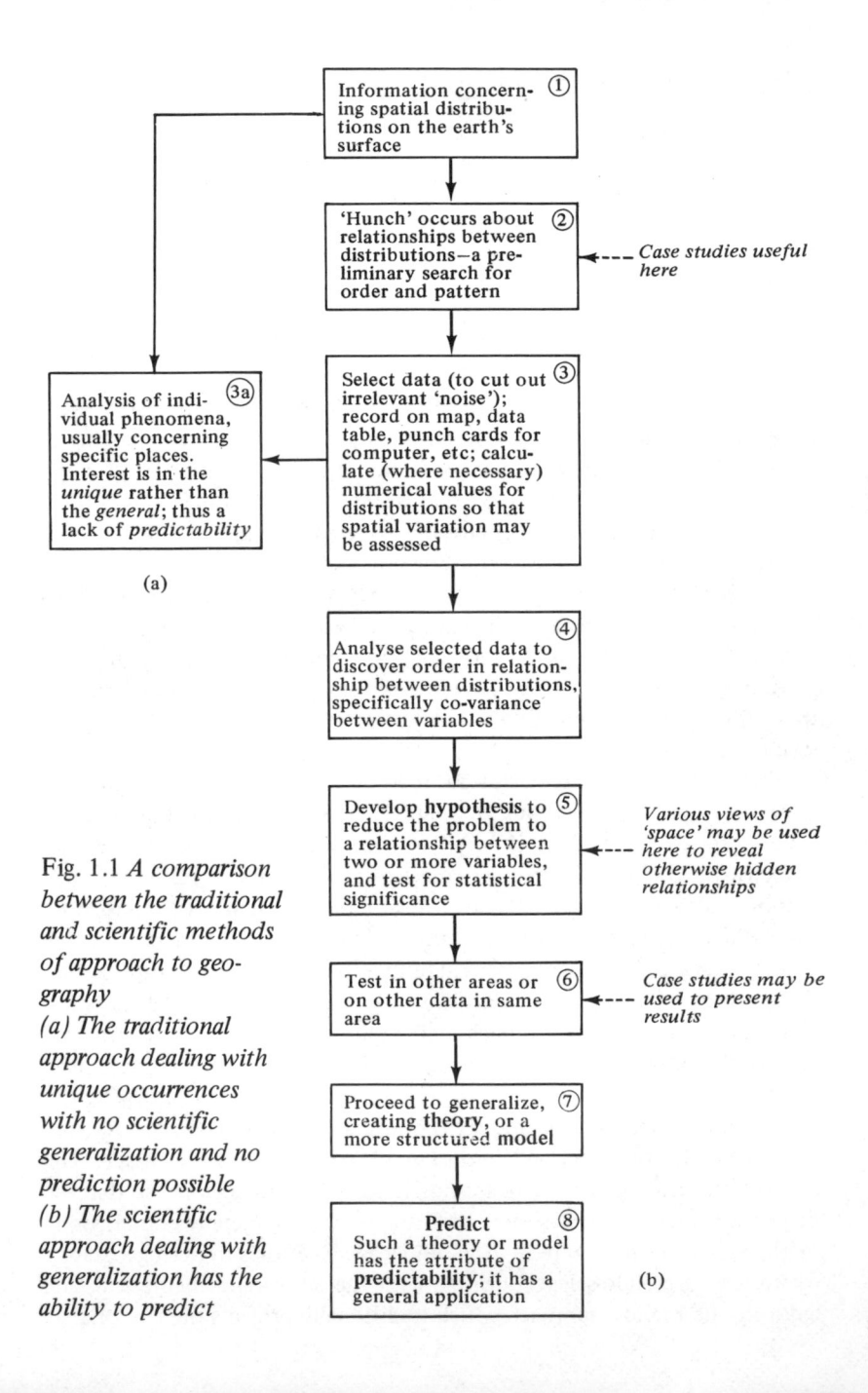

Fig. 1.1 *A comparison between the traditional and scientific methods of approach to geography*
(a) The traditional approach dealing with unique occurrences with no scientific generalization and no prediction possible
(b) The scientific approach dealing with generalization has the ability to predict

ter of place or region. Critics say that concentration on statistical techniques causes the geographer to ignore the earth's surface, and thus the real subject matter of geography.

I do not wish to decry the importance of the literary side of geography, though works of true literary merit are rare in any branch of geography; nor is it my intention to make less of the need to give an understanding of place or region. I feel, however, that recent developments in the subject are placing geography firmly with the experimental sciences, although some consider it still has an important part to play as one of the humanities.

This change of emphasis is important, for as a science matures it leaves behind its concern with classification of unique objects and with simplified cause and effect relationships, and progresses on to the realm of generalization. Once generalizations can be made, as we shall see, theories may be developed and **prediction** becomes possible. Through the subject's ability to predict it becomes far more valuable to society, and so geographers are more likely to participate in decision-making in the important areas of planning and development.

The development of geography

In *Explanation in Geography*,[1] David Harvey seeks to clarify the way in which geographers have tried to explain distribution on the earth's surface. He summarizes these briefly (pp. 78–83) under the following headings:
(a) *Cognitive description*, which collects, orders, and classifies data; an example of this is the simple description and ordering of climatic data for an individual area.
(b) *Morphometric analysis,* which studies the shape and form of geographical patterns. An example of this approach is the analysis of transport networks. This is becoming an important part of a geographer's analyses as it does allow for:
(c) *Cause and effect analysis,* which was the nineteenth century approach to geography. Harvey suggests that the search for geographical 'factors' that govern distributions is a good example of the 'restrained use' of this type of analysis at the present time. An explanation of wheat growing in East Anglia as a result of certain climatic characteristics, relief, soil type, and size of fields may be quoted as being an instance of this approach.
(d) *Temporal modes of explanation,* a short step from (c), as it traces back cause and effect explanations over a long period of time. An

historical analysis of the development of parish and village characteristics in the Midlands of England could be used to illustrate this
approach.

(e) *Functional and ecological analysis,* which attempts to see phenomena
in terms of the role they play within a particular organization. Thus,
as Harvey says, towns may be analysed in terms of the function they
perform within an economy. Many geographers, not the least being
Stoddart,[16] feel that such studies form an important basis for their
work.

(f) *Systems analysis,* which provides a framework for studying the *whole*
structure of a society or an organization as a system of inter-locking
and interacting parts. Much of our work on conservation, pollution,
and urban development within society could be studied in this manner.

Harvey suggests that these forms of explanation in geography may be
paralleled by the following questions:

(a) How may the phenomena being studied be ordered and grouped?

(b) How are the phenomena organized in terms of their spatial structure
and form?

(c) How did the phenomena originate and develop?

(d) How were the phenomena caused?

(e) How do particular phenomena relate to and interact with phenomena
in general?

(f) How are phenomena organized as a coherent system?

These six methods of explanation are all part of the subject of geography,
and many can be seen to overlap. In the historical development of geography there has been a general move from *cognitive description* and
cause and effect analysis through *temporal modes of explanation* to
functional and ecological analysis and more recently to *systems analysis.*
Morphometric analysis is a method of explanation which is becoming more
important, because the relating of patterns of distributions on the earth's
surface to geometry does allow for a degree of prediction, which is, as we
shall see, of fundamental importance to a science.

These six methods of explanation in geography partly reflect the development of the discipline, and are worth keeping in mind when we analyse
our own approaches to it. Although a detailed description of the development of geographical thought would be out of place here, it is, perhaps,
worth tracing one or two of the more important strands.

So-called 'classical geography' developed strongly throughout the nineteenth century and was based largely on the collection, ordering, and
classifying of data, and on the analysis of Newtonian 'cause and effect'

relationships. Much of this geography has been called **deterministic**. It leaned very heavily on the role of the physical environment in shaping, or determining, man's activities. In this way geography, later in time, paralleled a similar stage in the development of other sciences.

During the latter part of the nineteenth century, and even, perhaps, to the present day, the development of geography has been very much guided by Darwinian thought. Fact-collecting, cataloguing, and classifying, as well as literary description with some cause and effect analysis, were all important results of man's continued exploration of the earth. But Darwin, through his work on evolution (particularly after the publication of *On the Origin of Species* in 1859), had an even greater and longer-lasting effect on geography.

D.R. Stoddart,[2] discusses Darwin's effect on the course of geography in detail. He points out that Darwin made four important points, and that while three of them directly impinged on geographical thought, the fourth, of great importance, was strangely neglected.

In summary the four points were:
1. The concept of development through time.
2. The concept of the relationship between an organism and its habitat.
3. The concept of selection and struggle.
4. The concept of random variation.

(1) *The concept of development through time*

Here Darwin's impact was largely through the idea that a population of organisms would evolve over a period of time. This was taken up strongly in the work of W.M. Davis, particularly in the idea of the evolution of rivers and landscapes through stages of youth, maturity, and old age. Elsewhere we find the idea of the evolution of soils, climax vegetation, patterns of settlement in historical geography, and more recently of towns and cities. This approach is the 'temporal mode of explanation' of Harvey.

(2) *The concept of the relationship between an organism and its habitat*

Stoddart quotes Darwin as saying: 'How infinitely complex and close fitting are the mutual relations of all organic beings to each other and to the physical conditions of life.'

This is, of course, ecology, or the analysis of organisms in terms of the role they play in their particular environment. Ecology became very much a part of the development of regional geography. The close relations between all organic beings and the physical conditions of life constitute an 'organic unity'. The region is thought of as having a degree of organic unity which one is encouraged to look for in explaining the very large number of individual phenomena on the earth's surface. Thus a wide selection of the various systematic branches of the subject—climate, vegetation, soils, agriculture, industry, etc.—were seen to be unified as a whole in the study of a region. This organic unity was used in an attempt to bring together the disparate elements of regional geography. More recently, of course, the ecological explanation in geography has again become important, but without the strong emphasis on man-land relationships.

In the early twentieth-century writings of Vidal de la Blache we see the man-land relationship, as an organic unity, developed to its greatest extent in the *pays* concept; man's activities in pre-Industrial Revolution France are shown to be very much a product of local regional variations in climate, soils, and so on. We see this also in the regional textbooks by such authors as J.F. Unstead and F.J. Monkhouse, who in effect follow upon A.J. Herbertson's development of the concept of the 'natural region'. The idea or concept of a natural region is one which regards a particular area of the earth's surface as possessing a degree of homogeneity or 'sameness' throughout, in terms of the results of the actions of a variety of geographical factors such as climate, relief, and economic policy. Even today—correctly so in the eyes of many—this still leads to the wish to understand the 'feeling' for a place.

This view is emphasized by the Sixth-Form and University Standing Committee of the Geographical Association which says, 'We need . . . to build up a satisfactory picture of what the country that is being studied is like, arousing . . . a feeling for its people and an appreciation of the problems they face.'[3]

On the whole this approach produces a geography more concerned with *unique* (non-general) phenomena and with *individual* places. As such it is non-scientific and relies greatly on its factual content and on the standard of writing for its interest.

(3) *The concept of selection and struggle*

This concept emphasizes the survival of the fittest, and was closely associated with the development of nineteenth-century *laissez-faire* economics and politics. It had its effect on geography also. Ratzel, consistent with this line of thinking, formulated his seven laws for the growth of states: 'Just as the struggle for existence in the plant and animal world always centres about a matter of space, so the conflicts of nations are in great part only struggles for territory.'[4]

The concept which came to be known as 'the struggle for *Lebensraum*' was further developed between the wars, and became closely associated with the expansionist aims of Germany in the 1930s. This organic view of political geography has long since dropped into disfavour, although this does not detract from the scholarly merit of Ratzel's writings.

(4) *The concept of random variation*

Darwin referred to the operation of chance in nature to produce a random variation in the characteristics of any one group of organisms. It is upon this randomly variable distribution within a population that the processes of natural selection operate.

It is surprising that geographers latched onto the *fact* of evolution, (Point 1) but ignored the manner in which it worked. This may have been in part due to the fact that Darwin himself abandoned the issue of random variation at a later date.

Only recently, therefore, have chance processes been recognized in geography, a century after the kinetic theory of gases was developed. In the physical sciences, Peter Haggett states, 'laws' are 'not deterministic but only statistical approximations of very high probability based on immense—but finite—populations.'[5]

Supporting this, Professor Bronowski has said, 'a society moves under material pressure like a stream of gas; and on the average its individuals obey the pressure, but at any instance, any individual may, like an atom of gas, be moving across or against the stream.'[6]

Thus in a science such as physics, *laws*, which may be used for prediction, are only *theories* working on probabilities of a very high order.

Generalization and the ability to predict are the very hub of 'scientific method' in geography. Even in 1905 the importance of prediction was recognized in the physical sciences: 'What is a good experiment? It is that which teaches us something more than an isolated fact. It is that which enables us to predict, and to generalize. Without generalization prediction is impossible.'[7]

It might be worth pointing out that by 'experiment' we mean the testing of a theory or a hypothesis (see p. 9).

Thus it would seem that the scientist is interested in the development of laws of general application, while the student of the humanities is more concerned with the nature of unique occurrences. Professor Fisher, referring to sixth-form geography students, said that 'a majority of the scientists among them are far more interested in systematic studies, working towards laws of general validity, than in the study of particular places, with which, by definition, the regional geographer is concerned.'[8]

He then quotes the dicta that: 'In all sciences we are being progressively relieved of the burden of singular instances, the tyranny of the particular', and 'The factual burden of a science varies inversely with its degree of maturity',[9] to explain, in part, the move away from regional geography.

Positive support for this trend is found in W. Bunge's discussion: 'In my opinion geography is the science of locations. Regional geography classifies locations and theoretical geography predicts them. Even more broadly, science is the deadly enemy of uniqueness. As the masterful Schaefer taught us, generality is science's weapon in our unending reduction of uniqueness.'[10]

So far we have looked all too briefly at what we might or might not consider to be a science, and how this bears upon the development of geographical thought. It is worth reaffirming that we are not passing value judgements on non-scientific geography. However, if we accept the statement that: 'Geography can be regarded as a science concerned with the rational development, and testing, of theories that explain and predict the spatial distribution and location of various characteristics on the surface of the earth',[11] then the development of theories and generalizations which may be tested or used in prediction are fundamental to the subject. It is to the procedure by which such theories are established that the term 'scientific method' is applied, and it is this procedure which we shall now attempt to analyse.

The scientific method

Look at Fig. 1.1; this shows the scientific approach to geography diagrammatically arranged, and compared with the more traditional approach. As you can see, the figure is drawn as a flow diagram, each box representing a step in scientific analysis, with the more traditional approach indicated as a side path.

Box 1 represents the starting point of all geographical study. This involves the pattern of spatial distributions (of man's activities particularly) on the earth's surface.

The infinite mass of information which is available to the geographer cannot be encompassed in one all-embracing study. Even if one divides the earth up into a whole set of regions, the study of the smallest in this manner would prove impossible. Data are multiplying all the time and one may become engulfed in just gathering and storing all this information, with little more than some attempt at classification and perhaps some cause and effect analysis. To a certain degree this characterizes much of what we still do, particularly in regional geography.

To avoid becoming immersed in purposeless classification of a vast amount of information, the scientist has to approach his material with some idea of the problem he wishes to investigate. In other words he has a 'hunch' about it, supported perhaps by casual evidence that there is some relationship between one pattern of distribution (called a **variable**) and another. The relationship between the number of farms of various types and their distance from large towns in the area under study would be an example of this. Such an approach is indicated in box 2.

Armed with some idea of his problem the geographer will next wish to select the data he needs, discarding what is irrelevant (box 3). Such data must then be recorded in some way, perhaps in map form, or as a table of figures (a data matrix), or on punch cards for computer analysis, or simply, but perhaps less scientifically, as a piece of prose. As will be stressed in S.I.G. 2 (*Science in Geography*, Book 2), careful measurement—where measurement is necessary—is important. This may be the measurement of pebble size on a beach, or of the journey length of a set of commuters, or of the characteristics of a distribution of farms (analysed by nearest neighbour analysis). He will then be able to assess the variation in the variable concerned over a part of the earth's surface. This change is called **spatial variation**.

At this stage the geographer will begin to look for order and pattern in the distributions which he is investigating (box 4). Any co-variance that exists between the variables being studied may be subjectively described in some such form as 'there is a tendency for the larger, higher-rent paying, department stores to group near the centre of the city'. However, greater precision than this is normally required in descriptions of distributions. Some of the statistical tools (e.g. nearest neighbour analysis) that may be used are outlined in Chapter 2, and fully explained in S.I.G.s 2, 3, and 4.

The analysis so far will enable the geographer to arrive at the formulation of a relationship between two (or more) variables, for example, farming type and distance from large towns. Strictly speaking, analysis of the *type* of farming is often difficult, so it is more likely that the *intensity* of farming would be related to the distance from urban areas. Such an untested assertion of a relationship is usually referred to as **a hypothesis** (Box 5), although a number of geographers refer to this as 'theory'. A geographer may test a hypothesis statistically for **significance** in order to see to what extent he can judge the probability that the relationship he has found could *not* have occurred by chance. In S.I.G. 4 the role of the rejection of the *null hypothesis* (where it is first assumed that any co-variation between a pair of variables *is* just a matter of chance association) in accepting the hypothesis that there is a significant co-variation between a pair of variables is fully discussed, together with relevant examples.

The stated hypothesis, in this case, could normally take the form: *that there is a negative or inverse relationship between the intensity of land use and the distance from the centre of a city.*

(A negative or inverse relationship is present when one variable varies inversely with another, that is, as variable 'A' increases, so variable 'B' decreases, and vice versa.) It is important to realize that this is only a hypothesis, which could prove to be untrue and would then have to be discarded.

Fig. 1.2 illustrates how a hypothesis may be formulated. Fig. 1.2(a) shows the return per acre* that a farmer obtains from each of his fields while Fig. 1.2(b) shows the degree of slope for each field. In this example the geographer studying the distributions had developed a 'hunch' that there would be some relationship between slope and returns.

The two maps represent the data selected by the geographer as being relevant, and examination of them leads to the formulation of the hypothesis:

that there is a negative or inverse relationship between the value per acre of land use and the degree of slope.

The two variables in this case are:

(a)   the value of the land use—a **dependent variable** (that is one that is thought to be at least in part dependent on another), in this case dependent on:

(b)   the degree of slope of the land—the **independent variable** (that is one that is thought to control, at least partly, the dependent variable).

Once such a hypothesis has been formulated it may be further analysed by plotting the data on a graph: the value of land use as the y values and the degree of slope as the x values. It is usual to show the independent variable on the x-axis.

Fig. 1.3 shows the data from Figs. 1.2(a) and (b) plotted on a graph. The points do not lie on a straight line; if they did the relationship would be a perfect one. In a perfect relationship, where the points all lie on a straight line, any increase in one variable is associated with a proportionate increase (or decrease) in the other. The straight line which is placed to go

*Where data were collected before metrication, examples will quote Imperial units.

Fig. 1.2 *Stewart Farm: (a) capital letters indicate land use and numbers indicate return in £s per acre per field; (b) numbers indicate the degree of slope of land*

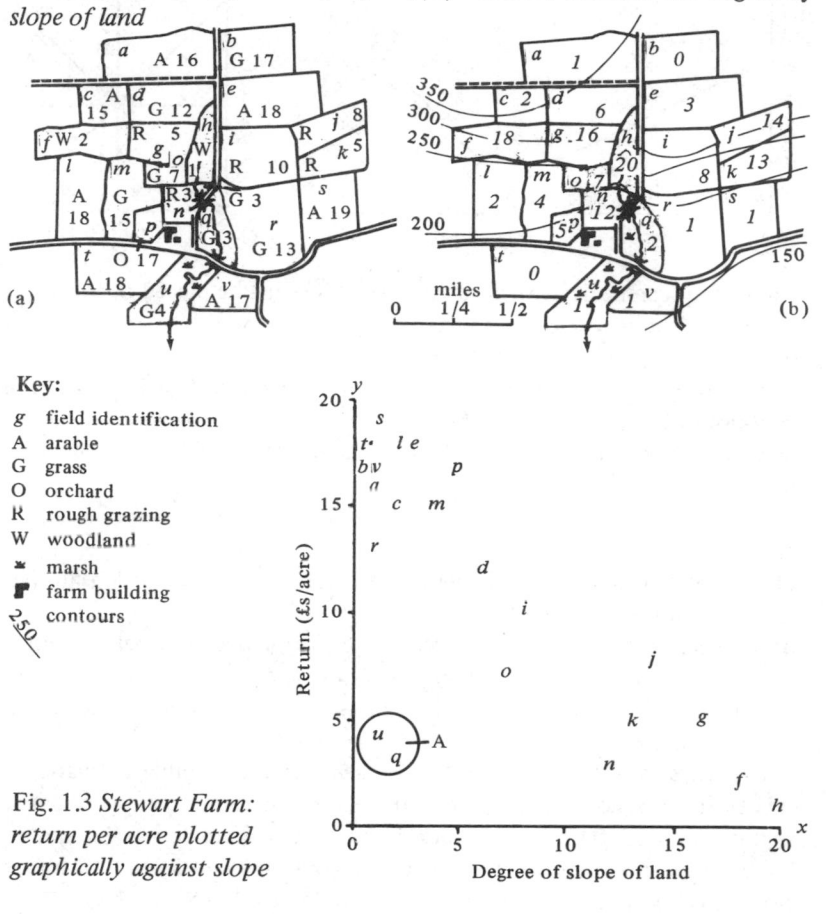

(a)

(b)

Key:

g	field identification
A	arable
G	grass
O	orchard
R	rough grazing
W	woodland
✳	marsh
⌑	farm building
	contours

Fig. 1.3 *Stewart Farm:
return per acre plotted
graphically against slope*

through the points as centrally as possible is called a **best-fit** line, and variations to either side of it indicate that other factors are at work to a lesser or greater extent (see particularly the group of points on the graph marked 'A'). In our example, the points are fairly closely grouped about such a line, and we can therefore suggest that there is a moderately close relationship between the two variables.

Fig. 1.4(a) shows a perfect relationship between two variables 'A' and 'B' while Fig. 1.4(b) shows a moderately good relationship between variables 'C' and 'D'. Fig. 1.4(c) on the other hand, shows a case where no

Fig. 1.4 *The degree of relationship between a pair of variables*

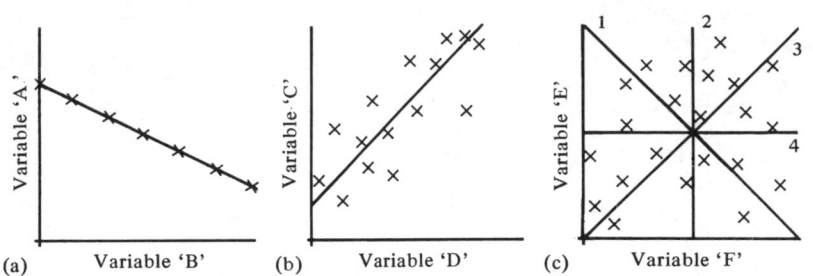

(a)   Variable 'B'   (b)   Variable 'D'   (c)   Variable 'F'

relationship can be discerned, as the best-fit line can be placed at any angle (1,2,3,4) without disclosing any tendency for the points to group along a line.

A relationship can be *negative* or *positive*. In the case of Fig.1.4(a) the relationship is a negative one—as 'B' increases so 'A' decreases in value. In the case of Fig. 1.4(b), the relationship is positive—as 'D' increases in value so does 'C'.

The act of fitting lines, as we will see in S.I.G. 4, is important because it will allow the geographer:

(a)   to give a definite mathematical description of the relationship (an equation for the line);

(b)   to describe accurately the closeness of fit (the degree of relationship);

(c)   to obtain an assessment of the significance of the relationship in terms of the probability of such a set of observations not occurring by chance.

The ways in which one may test for significance, formulate equations, and describe the closeness of fit are varied. Some are rather complicated and lengthy in computation, others are simple, but perhaps less exact, and do not cover all three aspects. They include regression analysis, and the use of correlation coefficients. These are dealt with in detail in S.I.G. 4.

At this juncture some comment about the group of points 'A' in Fig.1.3 should be made. These points are anomalous, not fitting into the general trend of readings, and, as such, indicate the operation of factors stronger than that of slope. Careful examination of the maps (Fig. 1.2) will reveal the reason for the very low values per acre despite the low slopes. As you can see, the low values appear to be associated with marshy land. Points such as these, not arranged close to the best-fit line, are known as **residuals** (see S.I.G. 4, Chapter 7), and the analysis of residuals, particularly from the point of view of *where* they occur, and the reasons for their occurrence, is of growing importance to the geographer.

Once the significance of a relationship can be established it can then be tested on data from elsewhere (box 6). If, through testing, the hypothesis is shown to have validity (general application), it can then be accepted as **theory**. Such development of theory is a process of generalization (box 7).

A theory—if it is a valid theory—can be 'proved' as it were, by applying it to data elsewhere. In other words it must have a general application, and not be tied to one particular, unique instance. If it has a general application, it may then be used for prediction.

Models

Many theories are, however, an integral part of a rather more structured representation of reality, normally referred to as a **model**. Haggett and Chorley have suggested that a model is 'a simplified structuring of reality which presents supposedly significant features or relationships in a generalized form' and that, 'as such they are valuable in obscuring incidental detail and in allowing fundamental aspects of reality to appear'.[12] Thus models pick out the important generalizations in the working of any specific geographical system, and (ideally) can be applied to all instances. Fig. 1.5 shows an example of such a model, in this case a simple one of urban structure, which can be considered as having general application to all towns. A study of the principles behind any particular model does facilitate an understanding of reality, and furthermore allows prediction to be made. In this way models are a vital part of planning.

Fig. 1.5 *The concentric model of urban structure*
Source: Park, R.E., Burgess, E.W., and McKenzie, R.D., *The City* (Chicago, 1967) pp. 51,53

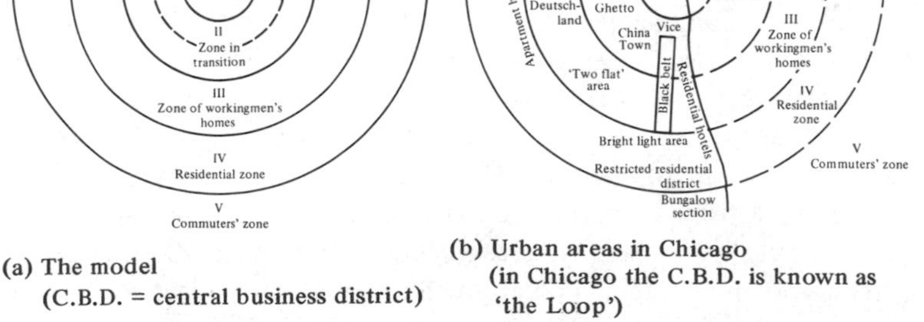

(a) The model
(C.B.D. = central business district)

(b) Urban areas in Chicago
(in Chicago the C.B.D. is known as 'the Loop')

Models can vary greatly in complexity: some may be so simplified that they carry no information of worth, while at the other extreme the model may be so complicated that it has no advantage over a description of all the complexities of reality, and no insight is conveyed.

The Burgess 'Ring' model shown in Fig. 1.5 is highly simplified, yet useful in dealing with the economics of land rents in a freely competitive situation. Competition for land accessible to a large market bids up the price of land, having the effect of sorting out those land uses that can best exploit the most central location, with rings of other land-use types arranged in order around. This model becomes less useful, however, when other aspects of urban geography are being studied. Fig. 1.6, for example, shows a slightly more complex model which may be used when the distribution of shopping centres within a city is being studied.

Fig. 1.6 *Model of shopping areas within a city*

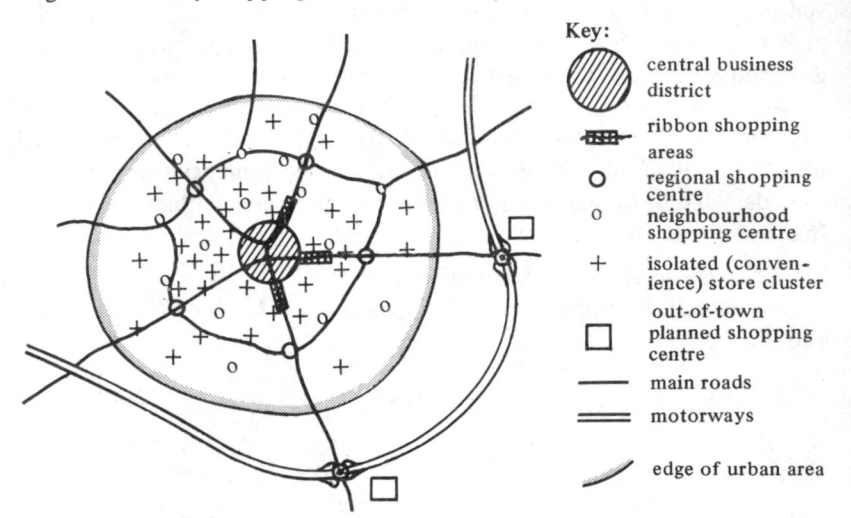

Key:

⊘ central business district

▦ ribbon shopping areas

◯ regional shopping centre

○ neighbourhood shopping centre

+ isolated (convenience) store cluster

▢ out-of-town planned shopping centre

— main roads

═ motorways

╱ edge of urban area

Thus more than one model may be required to make an adequate study of a subject such as urban geography, different 'structurings of reality' being required to 'carry' the essential information.

In this manner testable theories that allow prediction are developed. The testing of these theories may take the form of field work, or of practical work using secondary sources for data (e.g. census material). The sources for such work are discussed in some detail in S.I.G. 2. In carrying out field work or research, anomalies will undoubtedly appear, and any that are significant will be of utmost importance in developing new theory bringing in other factors.

Case studies

A comment on the role of *case studies* may be made here. Case studies are also referred to, in some quarters, as sample studies. Some people consider that this is an incorrect term, as it suggests that the individual farm or factory has been chosen randomly. Strictly speaking, however, this does not follow. Such a sample is a highly biased choice, based on its apparent usefulness. The whole subject of sampling is dealt with in S.I.G. 2, Chapter 3.

Case studies figure prominently at all levels in school geography. The intensive study of an individual village, farm, or factory has many useful aspects. A great deal may be learned, and some general conclusions may be drawn incidentally. However, the unthinking use of case studies has its drawbacks. Their use, without any background theory, is likely to create erroneous impressions of the working of *all* villages, farms, and factories. In these instances false generalizations are drawn In addition, much useless information, irrelevant to the topic under study, may be gathered.

There are two specific areas where case studies are important to geography. The first is where the analysis of, say, an individual farm can contribute to the understanding of the working of farms *as a system*, with input, throughput, and output, whose activities are in response to a whole variety of social and economic forces (see pp. 17–23).

The second area where case studies may be valuable is where the testing of a hypothesis or theory is being carried out. For the work to be truly scientific, the case studies must be properly sampled or chosen (see S.I.G. 2), and a satisfactory number be chosen for the significance of the results to be accepted (see S.I.G. 4, Chapter 3). In a teaching situation such an approach may well be impractical, as it would be if you, as a student, were developing a small thesis along these lines. In this case it may be acceptable to choose just one or two case studies to illustrate the working of the relevant theory. Such a limitation would not, however, be acceptable in truly scientific, original, geographic research.

Many models already exist in both human and physical geography. Some are introduced early in our geographical education. In meteorology, for example, we are all familiar with the frontal model of a depression. Until this had been developed earlier this century it was a very difficult task to make sense of the countless readings of temperature, humidity, pressure, wind, cloud, and so on, that were gathered from the many weather stations throughout Britain and elsewhere. With the development of such a model the observed characteristics of individual depressions could be compared with the model, and from then on there was a marked increase in the efficiency of prediction. Other models in physical geography may

Fig. 1.7 *Descriptive model of the features of glaciation*

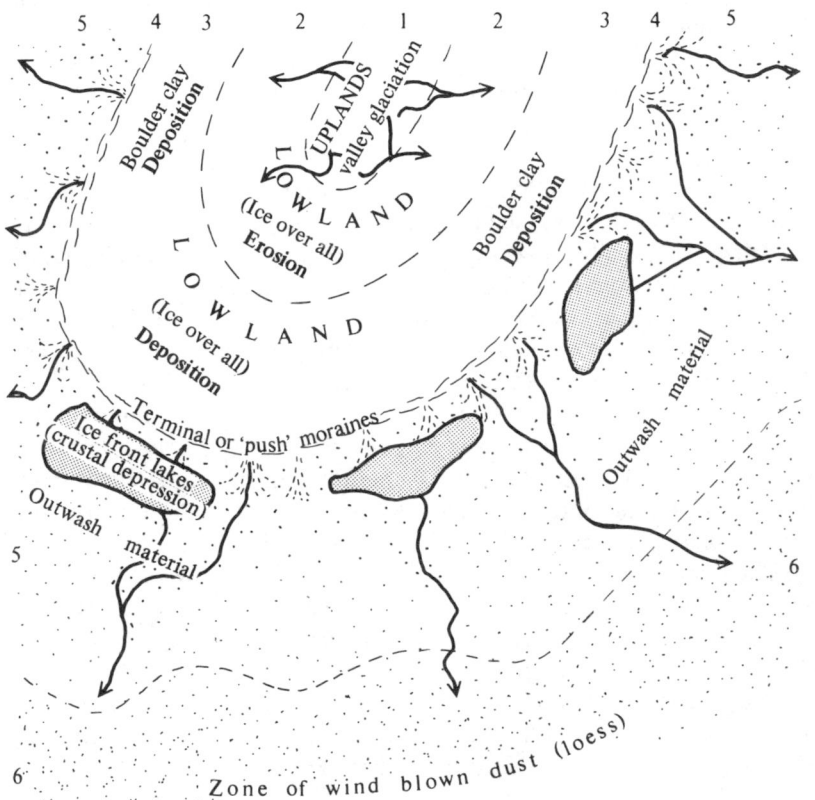

Key

Zone
1 Upland glaciation;
  erosional features dominant,
  overdeepened valleys, finger lakes, etc.
2 Lowland erosion:
  bare rock, sand and gravel deposits,
  ice gouged lakes
3 Lowland deposition:
  boulder clay, drumlins,
  kettle holes

Zone
4 Terminal or 'push' moraines
5 Outwash material:
  fans, deltas, and spreads of
  gravel, fluvial sorting, ice
  front lakes
6 Broad zone of wind blown
  dust (loess of China, limon
  of NW Europe). Permafrost
  features, solifluction

be largely descriptive (see the glaciation model, Fig. 1.7) or be more concerned with process and evolution, as in the Davisian cycle of erosion.

In human geography 'distance' is an important part of most models, although it may not be necessary for a model to include a distance element for it to be considered geographical. This is so, despite the obvious concern of the geographer for space. Fig. 1.10 gives such an example, which may be considered not only as economics, but also as geography.

Most of the models that do involve distance concentrate on the principle of **distance decay**. This principle is not solely a geographical one, and is met with in both the biological and physical sciences. It states that the *effect* of any action or set of actions that take place at a point will vary inversely with distance from that point. Thus in physics the strength of a magnetic field decreases inversely with the square of the distance from the magnet. This is an example of what is often referred to as the **inverse square law**. As similar principles apply to gravitational fields, the mathematical relationship is often referred to as the **gravity model**. Fig. 1.8 gives details of the gravity model. It also shows that a model can quite often be no more than a mathematical formula.

This decrease in intensity of activity over distance from a point (distance decay) is fundamental to a number of very important geographical models. This is not the place to go into the geographical theory behind such models, but a few may be mentioned. They include the various aspects of von Thünen's type of analysis (see Fig. 1.9), including the Burgess 'Ring' model of the structure of towns (see Fig. 1.5). The distance decay concept gave rise to the model known as Christaller's 'fixed k' system.

The gravity model (Fig. 1.8) has been 'lifted' complete from physics, and as such it is classified as an **analogue model**. The two elements of the formula—mass and distance—can be replaced by geographical parameters, for example the use of population of two settlements instead of the mass of the two bodies, and perhaps time instead of distance. In this way it might be possible to predict the flow of traffic between the two towns, or to calculate the **breaking point** between them. The breaking point is the point of indifference between fields of influence of two neighbouring towns. The two derived formulae—for interaction (in this case flow) between two towns, and for the breaking point—are given in Fig. 1.8.

Systems

One final aspect of scientific geography that deserves some mention is the way in which the various parts of the subject can be looked at as **systems** (not to be confused with the more usual study of geography

Fig. 1.8 *The gravity model*

$$I_{ij} = G\frac{M_i M_j}{d^2}$$

where: I_{ij} is the gravitational force between bodies i and j

M_i, M_j are the masses of the two bodies

d is the distance separating them

G is the gravitational constant

(a) The gravity model in physics

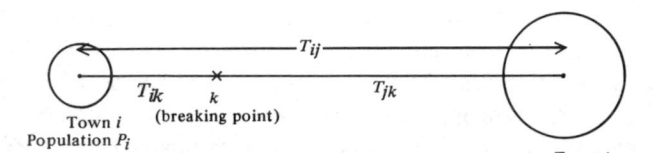

T_{ij}

Town i | T_{ik} | k | (breaking point) | T_{jk}
Population P_i

Town j
Population P_j

$$I_{ij} \propto \frac{P_i P_j}{(T_{ij})^\alpha}$$

where: I_{ij} is the expected interaction

P_i, P_j are the populations of the towns i and j

T_{ij} is the time taken between i and j for the means of transport being considered

α is an exponent related to the mobility of the population and may be varied to any value between about 1 and 3 — i.e. it must be approximately 2 in value — to suit conditions in, say Ghana or the U.S.A.

$\propto$ is the sign representing 'is proportional to'

(b) The gravity model in geography, using town populations and time distances

$$T_{jk} \propto \frac{T_{ij}}{1 + \sqrt{(P_i/P_j)}}$$

where: T_{jk} is the 'breaking point' distance from the larger town (j)

(c) The 'breaking point' formula derived from the gravity model

Fig. 1.9 *Von Thünen's system of land use (a) without complication of transport lines (b) with distortion due to navigable river (a line along which friction of distance is reduced) and a small city with its own production zones*
(**Source**: after Chisholm, M., *Rural Settlement and Land Use* (Hutchinson 1962), p.29. Chapter 2 in this book gives a very useful introduction to the importance of Von Thünen's work prior to 1826)

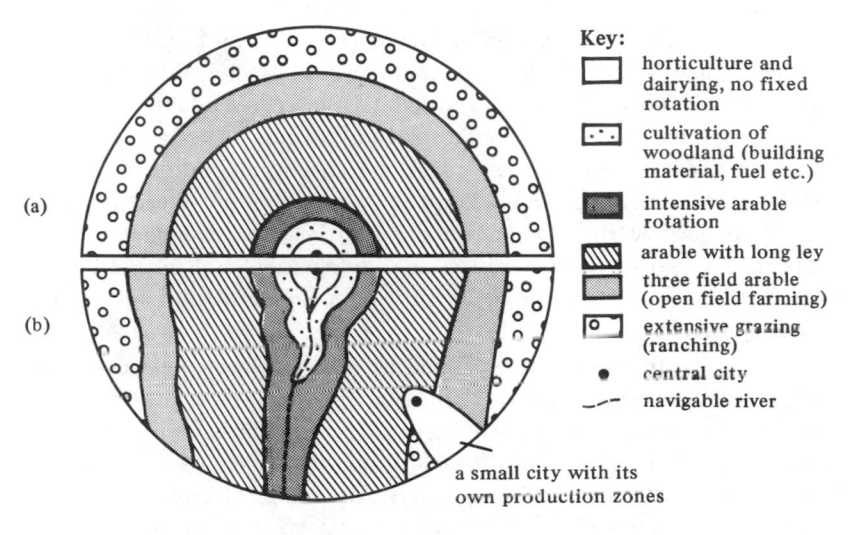

Key:

☐ horticulture and dairying, no fixed rotation

[∴] cultivation of woodland (building material, fuel etc.)

▨ intensive arable rotation

▧ arable with long ley

☐ three field arable (open field farming)

[○] extensive grazing (ranching)

● central city

–·– navigable river

(a)

(b)

a small city with its own production zones

through its various 'systematic' branches such as historical, physical, and economic.) Stoddart has described the system as a fundamental integrating concept in geography.

This approach is, perhaps, the answer to Professor Fisher's charge that 'geography is in serious danger . . . of over-extending its periphery at the expense of neglecting its base.'[13] Fisher suggests that 'a new school of geography, rejecting the common thread of thought which can be traced back from Stevens, via Mackinder, to Ratzel, and beguiled by the attractions of quantification, has focused its attention on ever more meticulous measurement, particularly of geomorphological and socio-economic data,' and goes on to say that 'the regional/ecological approach which provides the vital link between the physical and human aspects of the subject'[13] is neglected.

Professor Fisher is therefore making a plea for a return to a study of 'place' through regional geography, coupled with a greater investigation of ecological studies. I feel that few interested in a scientific approach to geography would deny the importance of the 'regional/ecological' approach in this context.

It is interesting to see that such work *is* being carried out by a geographer/ economist (Walter Isard, in America) on the study of regions. Here, however, the emphasis is being placed very much on taking a region defined economically, and studying the economic inputs and outputs. In other words such regional geography is being studied from a systems point of view.

In such a regional analysis in geography the whole working of a system, with its apparently static manifestations on the ground, the patterns of towns, villages, communications, and so on, is studied. In addition to this, a systems approach realizes the importance of *process* as well as *form*—that flows of goods, people, money, ideas, are an integral part of geographic analysis, as are also the various forms of input and output already referred to. These include such factors on the input side as government grants, foreign aid, money from exports, and immigration; and on the output side loans to other regions or countries, money spent on imports, and emigration.

A system can, perhaps, be defined as a set of objects related to each other through some form of circulation in which there are inputs of energy, circulations (or throughputs), and finally some form of output. Such a system normally has a visible form which over a short period of time remains more or less constant, but over a longer period of time may well show variations which are adjustments to changing inputs. Any change in input (energy) may result in change of form. This will continue until a state of **equilibrium** is again reached. Such equilibrium is usually referred to as **dynamic equilibrium**, and indicates that there is (theoretically) no change of form despite a continuance of the working of the system.

An obvious case of a system at work is the human body. There are daily inputs of energy (food) and outputs of waste matter and energy expended in various activities. Throughputs are varied, the most important perhaps being the activity of the blood system carrying oxygen and carbon dioxide, food in soluble form, hormonal secretions, and so on. From day to day we alter but little in appearance, although multitudes of cells have died and been replaced by others. Thus the body reaches a **steady state**, or dynamic equilibrium. Such self-regulation is referred to as **negative feedback**. Over a longer period, however, there are processes which bring about lasting changes in form, and so we have the period of adolescence, or the onset of old age. Progressive changes such as these are brought about by a process of **positive feedback**. The more disastrous interventions by man in, say, tropical agriculture have been of this type.

One final point concerns the body as a system. In the last paragraph we referred to the 'blood system'; this points to another attribute of many systems—they may well have smaller systems working within the main system. Such smaller systems are referred to as **sub-systems**.

The systems so far described are known as **open systems**, that is, they have inputs and outputs of energy and mass (or material). **Isolated systems** operate entirely within themselves without inputs or outputs of energy or mass, and are not met with in geography. In the case of **closed systems**, energy exchanges take place, but no transfer of materials occurs.

Haggett suggests that a world-wide study might be considered as a study of a closed system,[14] as important exchanges of energy (but not of material as yet) do take place between the sun, the earth, and outer space.

A more detailed account of systems may be found in *Systems Thinking*, edited by F.E. Emery, and in *Physical Geography: A Systems Approach* by R.J. Chorley and B.A. Kennedy.

Systems abound in all fields of geography. The main problem is the identification of the various elements. Systems diagrams are, in effect, only forms of models which define and trace the processes at work. Thus geographers may refer to the **urban system** when attempting to define the socio-economic forces at work in urban areas. In such a system there are inputs of energy in the form of income from trade, immigration, and perhaps government grants (see Development Area policy), for road improvement schemes, for health, welfare, and education. Outputs appear in the form of manufactured articles, products sold, out-migration of people. All these (and much else) are the result of socio-economic forces at work, and over a relatively long period of time we see far-reaching changes in form (e.g. new urban motorways, airports, new shopping precincts). These long-term changes are akin to the changes referred to in the case of the human body. Short-term changes, which do not alter the form or spatial pattern of the system, abound in the urban system. Cell destruction and replacement exists in the human system; in the urban system, minor road improvements (widening or resurfacing) and the refurbishing of sub-standard housing are examples of short-term changes.

The urban system may or may not be easily defined, but within its working one can distinguish a host of sub-systems, all of which knit together to produce the whole. These sub-systems are *sub-systems* when we are considering urban geography as a whole, but are also *systems* in their own right when we consider them individually. For example, each factory, home, school, or hospital may be considered as a system in its own right. Each

Fig. 1.10 *Simplified systems model for a nation's economy*

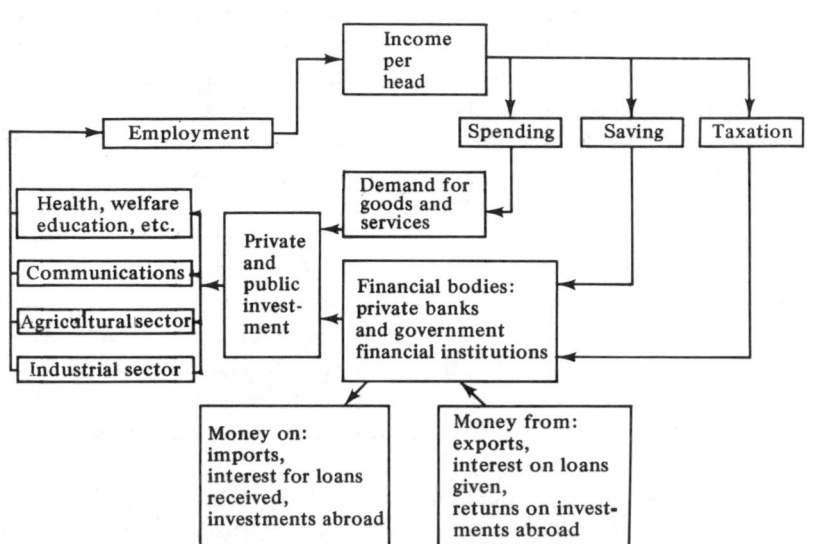

has its own inputs, throughputs, and outputs, but all these, and others, taken together make up the urban system. Some of their inputs and outputs are also those of the whole urban system, others can simply be considered as energy exchanges between one sub-system and another within the working of the whole urban system.

It is not a big step to consider the urban system as being a part of a larger country-wide economic system. As such, the urban system is itself a sub-system of this larger unit. Indeed a country's economic system is only a sub-system of the world economic system.

At this point the introduction of the word 'economic' rather than 'geographical' is questionable. I have used the term 'economic' here both because it is the more accepted term, and also because the system or model may be given in a non-spatial form. In the case of a country's economic system the geographer is particularly interested in the spatial implications. Fig. 1.10 shows a simplified systems model for a nation's economy: it is not a model taking space into account, and **the geographer should take one further step to study the spatial implications when the system is applied to any individual country.**

R.J. Chorley says* that an open system has some, or all, of the following

*The summary of Chorley's work is based on a summary made by Haggett.[15]

characteristics:

(a)   The need for an energy supply to maintain and preserve the operation of the system.

(b)   The capacity to attain a steady state in which changes in intake and output of energy and material are met by adjustments in form.

(c)   The ability to regulate itself by homeostatic adjustments. This is a negative feedback mechanism where the increase in an input is met by a form adjustment which lessens the input, and a return to the original state occurs. Governors on steam engines regulate steam pressure, and hence stop the system from 'running wild'.

(d)   The ability to maintain optimum magnitudes over periods of time. This implies that over a period of time when inputs are constant, the optimum level of the form and functioning of the system will also remain constant. This must, of course, allow for an inevitable time lag in adjustment after a period of changing inputs.

(e)   The ability to maintain its organization and form over time rather than tending towards maximum **entropy**. This is an important distinguishing characteristic of open systems. Maximum entropy occurs in closed systems and involves the 'running down' of the system over a period of time if there are no inputs or outputs of energy. A clockwork engine may show a trend towards maximum entropy. If it is not wound up, its system runs to a halt.

(f)   That they behave **equifinally** in the sense that different initial conditions may lead to similar end results. All Western urban systems appear to have similar form despite a great variety of initial conditions. In this way, urban systems may be said to behave equifinally. Such equifinal behaviour of geographical systems is very much the root cause of the general applicability of geographical models.

As Stoddart says: 'Systems analysis at last provides geography with a unifying methodology, and using it, geography no longer stands apart from the mainstream of scientific progress.'[16]

With this thought in mind we can close this section with a quotation from Professor Haggett: 'In the long run the quality of geography in this century will be judged less by its sophisticated techniques or its exhaustive detail, than by the strength of its logical reasoning.'[17]

References

1. Harvey, D., *Explanation in Geography* (Arnold, 1969), pp.78–83.
2. Stoddart, D.R., 'Darwin's Impact on Geography', *Annals of the*

Association of American Geographers, Vol. 56, no.4 (December, 1966), p.688, quotes C.R. Darwin.

3. Sixth Form and University Standing Committee of the G.A., 'Regional Geography in the Sixth Form Course', *Geography*, 56, no.3.

4. Stoddart, D.R. *op.cit.*, p.694, quotes F. Ratzel.

5. Haggett, P., *Locational Analysis in Human Geography* (Arnold, 1965), p.26.

6. Bronowski, J., *The Common Sense of Science* (Penguin, 1968), p.93.

7. Poincaré, H., *Science and Hypothesis* (Walter Scott, 1905), p.142.

8. Fisher, C.A., 'Whither Regional Geography?', *Geography*, 55, no.4. (November, 1970), p.376.

9. Fisher, C.A., *op.cit.*, quotes Prof. P.B. Medawar.

10. Bunge, W., 'Locations are not Unique', *Annals of the Association of American Geographers*, 56, no.2 (June, 1966), p.376.

11. Yeates, M.H., *An Introduction to Quantitative Analysis in Economic Geography* (McGraw-Hill, 1968), p.1.

12. Chorley, R.J., and Haggett, P., 'Models, Paradigms and the New Geography', in Chorley, R.J., and Haggett, P. (eds.), *Models in Geography* (Methuen, 1967), p.22.

13. Fisher, C.A., *op.cit.*, pp.374, 387.

14. Haggett, P., *op.cit.*, p.17.

15. Haggett, P., *op.cit.*, pp.18–19.

16. Stoddart, D.R., 'Organism and Ecosystem as Geographical Models', in Chorley, R.J., and Haggett, P. (eds.), *Models in Geography* (Methuen, 1967), p.538.

17. Haggett, P., *op.cit.*, p.310.

Further reading for Chapter 1

Abler, R., Adams, J.S., and Gould, P., *Spatial Organization* (Prentice-Hall, 1971), especially Chapters 1 and 2.

Bunge, W., *Theoretical Geography*, Lund Studies in Geography (Gleerup, 1966), especially Chapters 1, 7, 8 and 9.

Chorley, R.J., and Haggett, P., (eds.), *Models in Geography* (Methuen, 1967), especially Chapter 1.

Cole, J.P., and King, C.A.M., *Quantitative Geography* (Wiley, 1968), especially Chapters 1, 11, and 12.

Haggett, P., *Locational Analysis in Human Geography* (Arnold, 1965), especially Chapters 1, 9, and 10.

Yeates, M.H., *An Introduction to Quantitative Analysis in Economic Geography* (McGraw-Hill, 1968), especially Chapter 1.

Chapter 2

The use of mathematics in geography

This chapter deals mainly with how mathematics may be used in the new geographical methodology which was outlined in Chapter 1. It is important to point out that mathematics provides geographers (as any scientists) with a tool, a tool to produce a better understanding of the spatial relationships and working of the real world. It is a tool that serves to give greater precision to our arguments, a tool that helps to find and describe those relationships which might otherwise be difficult to uncover because of the confusing welter of factual information that assails our geographical senses.

Geography is very much at an emergent stage, discovering that many existing areas of mathematics (statistics, graph theory, topology) are of great use in building up the general theory that is the very essence of a mature science. Much that is now used in mathematics was developed as a result of man's needs and wishes over a long period of time. In this way, during the period of growth of physical science, mathematics was induced to develop along certain lines. More recently the needs of economic theory have inspired new lines of mathematical thinking, and now we are beginning to see how geography and the social sciences encourage new branches of mathematics to serve them.

Fig. 1.1 shows the scientific approach to geography. Mathematics, in the broadest sense, enters into such an approach at a variety of stages as set out below, where the relevant mathematical techniques have been listed under four headings:

(A)	Sampling	Fig. 1.1, box 3
(B)	Description of spatial distributions	Fig. 1.1, box 3
(C)	Correlation	Fig. 1.1, boxes 3 and 4
(D)	Topology and transformation	Fig. 1.1; not shown explicitly but does indicate variations in our perception of space and time.

Sections (A) to (C) more usually come under the specific heading of spatial statistics, and as such they form the detail of S.I.G. 3 and S.I.G. 4. This chapter is intended to give only an overview of the possible uses of mathematics in geography, and detailed explanations of method, application, and interpretation of these techniques are not given. Section (D), topology and transformation, is treated in this book in Chapter 3.

It is important to note that the scientific method may use a variety of statistics. At one end of the scale the statistics may be very detailed and complicated; at the other the assesment of quantity, degree of relationship, and significance may be very general and not make use of statistical formulations at all. At an elementary level, the concept of correlation may be presented only as a visual comparison between two lists of values of two variables (e.g., the value of the land use, and the distance of the land used from the farm buildings.) This may or may not be called 'statistical', but certainly the introduction of the concept of correlation is central to the scientific approach, with or without the use of more complicated techniques.

Sampling

One of the first areas in which a knowledge of mathematics may be used is that where the selection of data is made from what the geographer *perceives* as the real world. This selection follows on from the development of the geographer's 'hunch' about his problem, and is in part a response to the need to distinguish the relevant from the irrelevant (see Fig. 1.1, box 3). Even when the particular data required has been identified, the quantity of information may still be so large that it is necessary to select what one hopes will be a sample representative of the total information available. In other words it is desired, through sampling, to obtain **a set of data of the smallest size that is representative of the total population (all cases) or of the whole area being studied, and which is within a desired degree of reliability**.

The kinds of exercises involving sampling are dealt with in more detail in S.I.G. 2, Chapters 2 and 4, and S.I.G. 4, Chapter 5. Sampling could be used in studies such as the collection of data about channel and valley-slope characteristics along a stream, or the characteristics of the catchment area of a school of 2000 pupils, or the nature of the variations in the land use within a defined area.

In all these cases making use of the total information available would raise problems of time, effort, and cost. A reliable sampling method should result in a considerable saving in all three.

Description of spatial distributions

Once the data for our problem (Fig. 1.1, box 2) have been collected and recorded in some way (on maps, or in a data matrix, for example), as indicated in Fig. 1.1, box 3, it is necessary to describe these with some degree of precision so that they may be represented numerically. This is done in the following three stages:

1. The calculation of a numerical quantity to represent a spatial distribution, if the distribution is not already in number form.

2. The more precise assessment of variations in patterns of distribution in one area, or of values between one area and another. This provides us with a measurable quantity which varies in value. Such a quantity is called a **variable**, and varying as it does over the earth's surface it can be called a **spatial variable**.

3. The comparison, which is now possible, of such a variable with a selection of other spatial distributions which are thought to have some significance. The amount of agreement between the pairs of distributions is called co-variance and is important in expressing correlation (Fig. 1.1, box 4). The subject of correlation—an important part of scientific method—is dealt with below (p.40), and in more detail in S.I.G. 4, Chapter 6.

Fig. 2.1 *Contrasting farm distribution patterns*

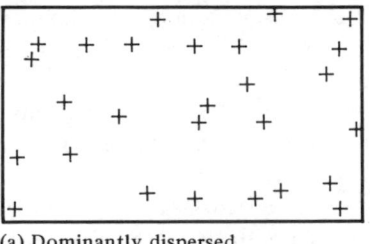
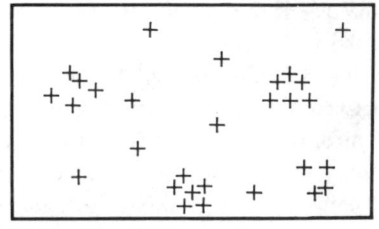

(a) Dominantly dispersed   (b) Dominantly clustered

An example of such a description is the comparison of the characteristics of the pattern made by farms in two separate areas, shown in Fig. 2.1. Some precision is required to describe these two apparently different patterns and this may best be done by using a statistic such as the **nearest neighbour coefficient** (see this book p. 33 and S.I.G. 3, Chapter 3), which, once stated, allows us more easily to relate the differences between the two patterns to variations in some other factors. This estimation of co-variance would be a part of stage 3. If, in this case, we find that variation in the pattern of farms exists in conjunction with, say, variation in rainfall pattern, we might have highlighted a significant relationship. If we find a significant relationship between settlement pattern and another distribution such as population density, rainfall, or some other variable feature, then we may have revealed a factor contributing to the socio-economic organization of the area.

A second example of this approach would be the use of the data given in Figs. 1.2(a) and (b). Here the two variables are value of land use and the degree of slope of land. Both variables are already in numerical form (e.g. £15 per acre, 3° of slope), so that stage 1 is already accomplished. The values of each of these two variables vary over the part of the earth's surface covered by the farm. The individual fields represent the areas which vary in value between one part of the farm and another (see stage 2). The two sets of data may be analysed to discover the *degree* of relationship between them, and to discover its significance.

Thus if we wish to analyse a distribution pattern scientifically, we must attempt to change the pattern into some numerical quantity (stage 1), if it is not already in this form, before going on to stages 2 and 3. To do this we have a variety of statistics to describe point and line patterns, and the shape of areas in numerical form. Although we will look at the description of pattern in the next few pages we must first briefly consider the dimensions that normally concern the geographer.

Dimensionality

Before precise description and analysis can be carried out, it is necessary to understand something of the nature of the elements in a spatial distribution. Generally in geography we are dealing with three basic dimensions:
(1)   Points   (0 dimensions)
(2)   Lines   (1 dimension)
(3)   Areas   (2 dimensions),
but becoming of greater importance is the study of:
(4)   Surfaces (3 dimensions)
(5)   Time, which some consider to be analogous to the others.

The geographer, therefore, normally describes the actual spatial distribution of phenomena on the earth's surface in terms of the first three headings. Fig. 2.2, for example, shows the parish of Gristhorpe in Yorkshire. Here the farms are shown as a point distribution; the roads, lanes, and the parish boundary as line distributions; and the village itself as a two-dimensional 'patch' on the larger area of the parish. In this instance we are dealing with points, lines, and areas distributed in a pattern on the earth's surface.

Fig. 2.2 *The parish of Gristhorpe, Yorkshire: a pattern of points, lines and areas*

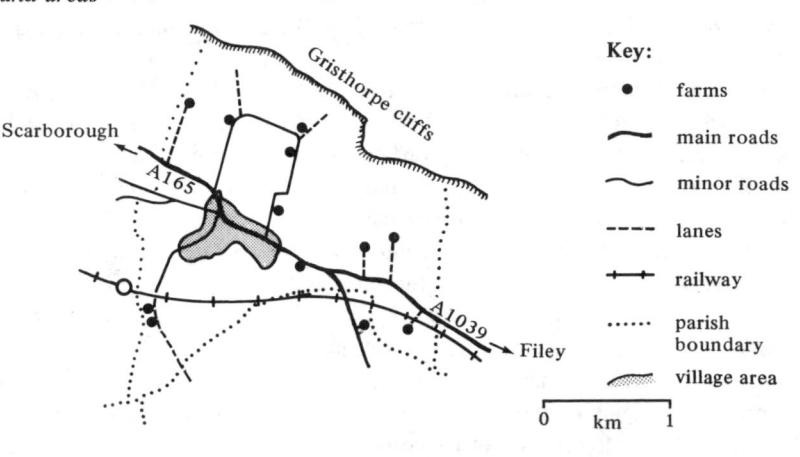

Scale is an important component. The farms in Fig. 2.2 are here considered as points, but if one were interested in the working of an individual farm, this being a larger scale, the farm buildings would be considered in two dimensions as areas, as shown in Fig. 2.3(a). In a similar manner, when one is interested in the distribution of buildings in the parish, the **area** of the village becomes a clustered set of **points** as shown in Fig. 2.3(b).

Thus both scale and the reason for study will affect the precise way in which the spatial distribution is represented.

The description of point and line patterns and of shape are dealt with in more detail than below in S.I.G. 3, Chapters 3 and 4.

(1) *Description of point patterns*

Point patterns may be described in two ways:

(a)   the degree of clustering: position on a scale ranging from completely clustered to completely ordered;

(b)   the measure of mean centre, and a measure of scatter about this point.

(a) *The degree of clustering.* It is possible to find the degree to which any point distribution departs from a purely random one, such as develops on dry paving-stones at the beginning of a rain shower. Point distributions can

Fig. 2.3 *The effect of scale and purpose of study on dimensionality.*
(a) Old Nick Farm. The study of a farm: at large scale, the farm is
represented in two dimensions (b) Gristhorpe. The study of the
patterns of farms in a parish: the farms are represented as points of
no dimension

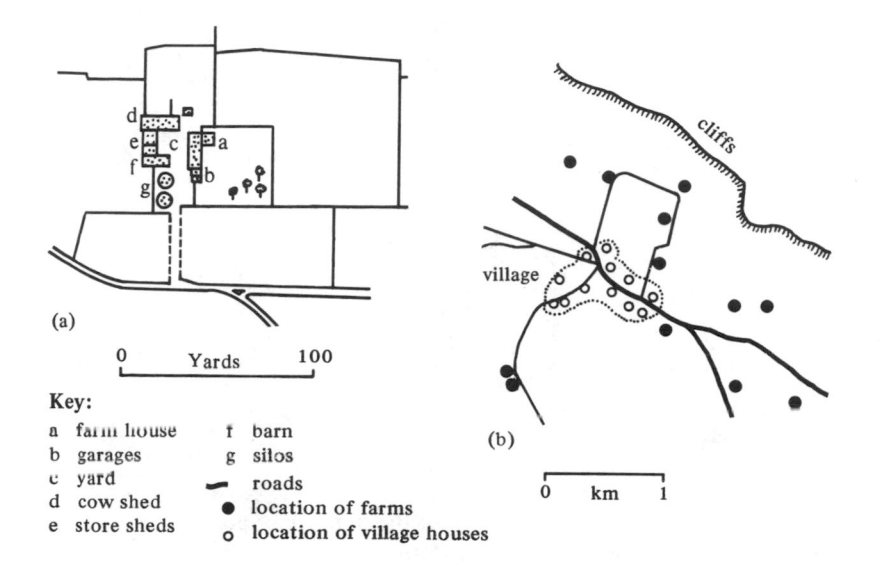

(a)

0   Yards   100

Key:

a	farm house	f	barn
b	garages	g	silos
c	yard	⚊	roads
d	cow shed	●	location of farms
e	store sheds	o	location of village houses

(b)

0   km   1

vary from one that shows all points arranged as far from each other as possible (a hexagonally ordered or completely anti-clustered distribution), through a so-called 'random' pattern, to one which is completely clustered (all are located on one point).

Fig. 2.4 shows three distributions of points: the perfectly ordered, the apparently random, and the highly-clustered.

Few geographical distributions are similar to an apparently random pattern; a variety of factors is likely either to have favoured clustering of activity (agglomeration, see Fig. 2.4(c)), or to have produced a scattered pattern that is an approach to efficient ordered spatial cover (Fig. 2.4(a)).

The degree of clustering is most easily found either:

(i) by use of chi-squared (χ^2)—which is also used in correlation and significance testing—see S.I.G. 3 and S.I.G. 4. , or

(ii) by use of nearest neighbour analysis (R_n)—see also S.I.G. 3.

(i) **Chi-squared.** In the case of χ^2, which is perhaps the less satisfactory of the two measures, we divide the area under consideration into a number

Fig. 2.4 *Three contrasting point patterns (a) perfectly ordered
(b) 'random' and (c) clustered*

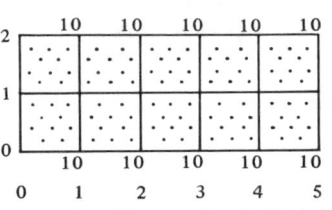

(a) 100 points placed in an ordered distribution (e.g. apple trees in an orchard)

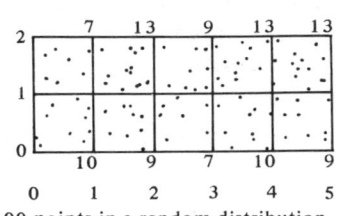

(b) 100 points in a random distribution
(e.g. raindrops on a pavement)

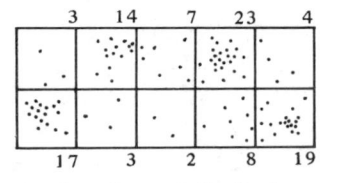

(c) 100 points in a clustered distribution
(e.g. houses in a rural area of the Midlands)

of equal-sized squares. We then count the number of points (farms, shops) in each. This is called the **observed frequency** (O) for each square. The total number of points is then divided by the number of squares to obtain the **expected frequency** (E). These two sets of values are related in the following formula:

$$\chi^2 = \sum \frac{(O - E)^2}{E}$$

where O is the observed frequency for each square,

  E is the expected frequency for each square,

  Σ is the instruction 'to sum' for all squares.

In the case of a perfectly ordered distribution, $\chi^2 = 0$. The higher the total obtained, the greater is the degree of clustering. The size of grid squares is very critical and χ^2 can only be used where two or more distributions are being compared *with each other*. The distributions must all be at the same scale and use must be made of identically sized grids. For comparative purposes, the method is satisfactory, but the absolute values for χ^2 mean little.

Chi-squared is of much greater value in testing the significance of the relationship or co-variance between two distributions (see S.I.G. 4).

(ii) **Nearest neighbour analysis.** This is a more useful, descriptive spatial statistic which relates the density of points in an area to the mean of the nearest neighbour distance. The nearest neighbour distance is the distance from a point in the distribution being studied to its nearest neighbouring point. The nearest neighbour statistic (R_n) is calculated using the following formula:

$$R_n = 2\bar{d}\sqrt{\frac{n}{A}}$$

where $\bar{d}$ is mean nearest neighbour distance for all points,

n is the number of points being considered,

A is the measurement of the area concerned.

R_n will always lie between 0·00 (for a completely clustered pattern) and 2·15 (for a completely ordered pattern). An R_n value of 2·15 is produced where the points are located in a hexagonal pattern at the vertices of equilateral triangles.

If a value of 1·00 is obtained then the pattern is often misleadingly called 'random'; the pattern that produces an R_n value of 1·00 may well be similar to the pattern made by a truly random occurrence, **but it need not be so.**

(b) *The calculation of mean centre and the measure of dispersion about this point.* The mean centre of a distribution of points may be broadly described as their 'centre of gravity'. Any point on a surface can be described in terms of both horizontal (x) and vertical (y) coordinates with reference to a **point of origin.** If a point of origin is assumed, as shown in Fig. 2.5, each point in the distribution has an x and a y coordinate. The mean centre of such a set of points can be calculated by finding the mean of all the xs $(\bar{x})$ and the mean of all the ys $(\bar{y})$. Such a point is denoted by $\bar{x}, \bar{y}$.

A measure of the dispersion of points around the mean centre is given by the **standard distance.** The standard distance (S.D.) is related to standard deviation in non-spatial statistics (see S.I.G. 3, Chapter 2) and is calculated by using the formula:

$$\text{S.D.} = \sqrt{\frac{\Sigma d^2}{N}}$$

where d is the distance of any individual point from the mean centre,

N is the number of points being considered,

Σ is the instruction to sum the d^2s.

In the case of a 'normal' distribution (e.g. the pattern made by a series of carefully aimed shots on a target, where most are close to, and only a

Fig. 2.5 *Relationship between one, two and three standard distances and the proportion of points covered. Note that each point can have its location described in terms of rectangular coordinates*

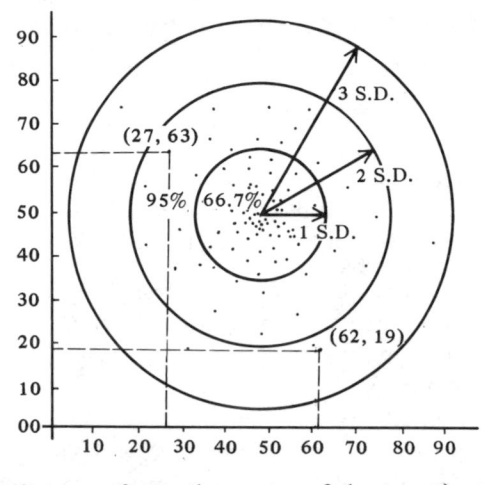

few at greater distances from, the centre of the target) approximately 67 per cent of the points should lie within a radius of one S.D., and 95 per cent within a radius of two S.D.s (see Fig. 2.5). If the points are not distributed normally (e.g. farms around the edge of a mountainous island) these percentages will not apply. Therefore variations in the standard distance, and the question of whether fewer or more than two-thirds of the points lie within one S.D. of the mean centre, can be used to describe the degree of dispersion of a set of points.

Detailed descriptions of both the mean centre and standard distance are to be found in S.I.G. 3, Chapter 3.

(2) *Description of line distributions (networks)*

Network analysis has become important in geography only in recent years. Until about two decades ago transport networks tended to develop in an apparently random manner, but the rationalization of railway systems, the building of motorways, and the development of other types of communication networks caused network analysis to be developed, if only for economic motives. When, for example, motorways cost £1 million per kilometre to build, it is of considerable importance that the form and extent of proposed network pattern is carefully considered.

The description of networks is valuable too in allowing a comparison between one country's network and another. This variation may then be related to variation in, for example, mean income per head, and thereby used as a possible indicator of economic development. Certainly the variation in the characteristics of networks is considered a reflection of certain spatial aspects of the socio-economic system.

Fig. 2.6 *A simple network*

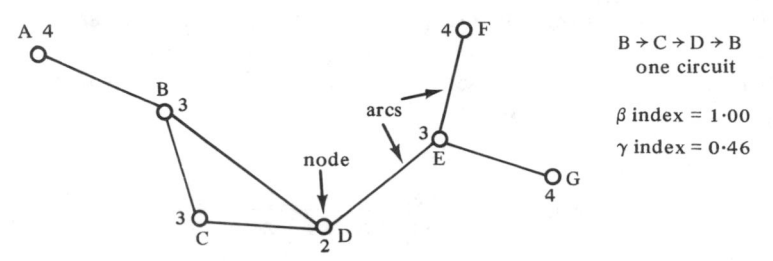

B → C → D → B
one circuit

β index = 1·00
γ index = 0·46

Numbers by nodes are König numbers. Most central node (lowest number) is D

Networks may be described in a number of ways, and reference to Fig. 2.6 will help explain how various indices may be constructed. At this point we will briefly consider the following:
(a)   the **connectivity** of a network
(b)   **centrality** within a network
(c)   the **spread** and **diameter** of a network
(d)   **detours**

(a) *The connectivity of a network.* **K.J.** Kansky (1963) developed a number of descriptive indices measuring the connectivity of networks. These included:
 (i) The **beta** (β) index,

$$\beta = \frac{\text{arcs}}{\text{nodes}} \qquad \text{(see Fig. 2.6)}$$

The beta index ranges from 0·0 for networks consisting of only nodes, and no arcs, through 1·0 and higher for networks that are well connected.

(ii) The **gamma** (γ) index,

$$\gamma = \frac{\text{arcs}}{3\,(\text{nodes} - 2)}$$

This is another index which describes numerically the connectivity of a network. This index will always lie between 0·00 and 1·00 (for a completely connected network).

(b) *Centrality within a network.* A useful index of centrality is the **König Number**. The König number of a node is taken as being the number of arcs by the shortest path to the node which is farthest from the node in question. König numbers have been placed on Fig. 2.6.

(c) *The spread and diameter of a network.* Kansky also examined the **diameter** of networks, which is taken as being the number of arcs on the shortest route between the two furthest points on the network.

Kansky developed two useful indices derived from diameter to measure the spread of the network:

(i) the Pi (π) index,

$$\text{where } \pi = \frac{\text{total mileage of network}}{\text{mileage of diameter}}$$

(ii) the Eta (η) index,

$$\text{where } \eta = \frac{\text{total mileage of network}}{\text{number of arcs}}$$

(d) *Detours.* The deflection of routes by physical and other barriers is an important geographical phenomenon, which can be measured by a detour index,

$$\text{where detour index} = \frac{\text{actual road (rail, etc.) distance}}{\text{straight line distance}} \times \frac{100}{1}$$

This index is useful in assessing the results of removing links from or adding links to any given network, or for assessing the degree of change brought about by new means of transport. An example of the latter would be the lessening effect of a hill barrier on communication as country roads give way to motorways and other forms of communication.

(3) *Description of shape*

The measurement of shape has always presented difficulties. Most indices are ratios involving the parameters given in Fig. 2.7, and are designed to lie between 0·0 and 1·0. Closeness to 0·0 indicates elongation, while a circle would have a ratio of 1·0. *If* shape can be measured satisfactorily, then any one of these shape indices varying between 0·0 and 1·0 may be useful for comparing shapes of areas such as parishes, and for relating variations in shape to other factors (e.g. geology). For further detail see S.I.G. 3, Chapter 4.

Five shape indices are given in section (3) of the summary of description of spatial distributions given below.

Fig. 2.7 *Basic parameters in measuring shape*

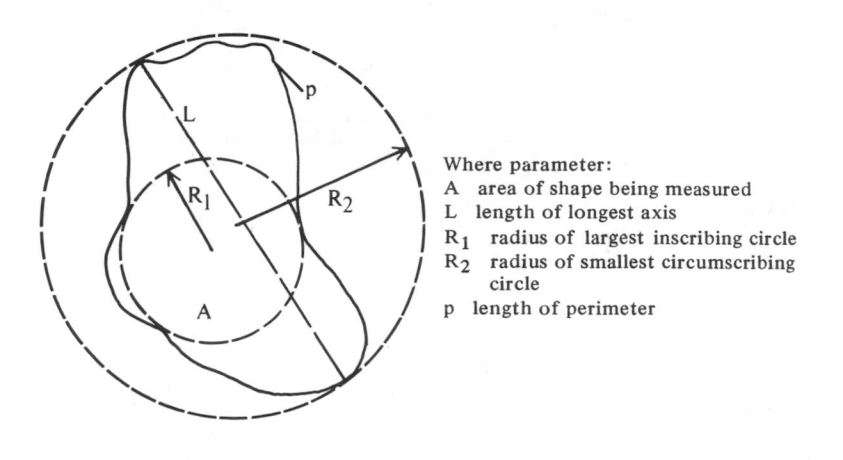

Where parameter:
A   area of shape being measured
L   length of longest axis
R_1 radius of largest inscribing circle
R_2 radius of smallest circumscribing circle
p   length of perimeter

Summary of description of spatial distributions

(1) Description of point distributions

    (a) Randomness   (i) Chi-squared (χ^2) $\chi^2 = \sum \dfrac{(O - F)^2}{E}$

                     (ii) Nearest neighbour coefficient (R_n) $R_n = 2\bar{d} \sqrt{\dfrac{n}{A}}$

    (b) Dispersion   (i) Mean centre $(\bar{x}, \bar{y})$

                     (ii) Scatter about mean centre $(S.D.)$ $S.D. = \sqrt{\dfrac{\Sigma d^2}{n}}$

(2) Description of networks

    (a) Connectivity   (i) Beta index (β) $\beta = \dfrac{\text{arcs}}{\text{nodes}}$

                     (ii) Gamma index (γ) $\gamma = \dfrac{\text{arcs}}{3\,(\text{nodes} - 2)}$

    (b) Centrality   König number—of any node is the number of arcs by the shortest path available to the node furthest from the one in question.

    (c) Diameter   (i) Number of arcs on the shortest route between the farthest points on the network.

(ii) Spread of
      network (π) $\pi = \dfrac{c}{d} \left(\dfrac{\text{total network mileage}}{\text{mileage of diameter}}\right)$

(η) $\eta = \dfrac{\text{total network mileage}}{\text{number of arcs}}$

(d) Detour index $\dfrac{\text{actual road distance}}{\text{straight line distance}} \times \dfrac{100}{1}$

(3) Description of shape

(i) $S_1 = \dfrac{A}{0 \cdot 282\, p}$ (iv) $S_4 = \dfrac{A}{\pi(0 \cdot 5\, L)^2}$

(ii) $S_2 = \dfrac{A}{0 \cdot 866\, L}$ (v) $S_5 = \dfrac{1 \cdot 27\, A}{L^2}$

(iii) $S_3 = \dfrac{R_1}{R_2}$

See Figs. 2.6 and 2.7 for keys to symbols.

(4) *Surfaces in geography*

It is worth pointing out here that traditionally the geographer has had little interest in the third dimension. Of course he has been interested in physical relief, but this is usually reduced to a two-dimensional map using isolines (in this case contours of altitude). Other physical and human data have also been represented in two dimensions where three-dimensional model representations might have been made. The difficulties in making three-dimensional models of such distributions as mean annual rainfall and population density offset their usefulness, although, of course, drawings of these distributions in three dimensions help considerably in making qualitative assessments. Computer programmes may also now be made to produce three-dimensional drawings or images on a screen from any angle, although the actual data that is stored in the computer is of greater import-ance in the carrying out of further work.

Correlations between distributions are much easier to make when represented in two dimensions so that it is not surprising that geographers have been mainly concerned with what can be graphically expressed on a sheet of paper. Nevertheless it is important to remember that distributions of points, lines, and areas can occur in three dimensions. As you can see in Fig. 2.8(b), the three-dimensional shells contain their own distributions of points, lines, and areas (although these latter are not specifically shown).

Fig. 2.8 *Essential ingredients of the urban environment: (a) part of a fictitious town; (b) the buildings as hollow containers (wardrobes or chest of drawers); (c) the channels designed primarily for movement; (d) the smaller mobile objects using the channels (c) to move between the containers (b)* (**Source**: Cole, J.P., and King, C.A.M., *Quantitative Geography*, p.399)

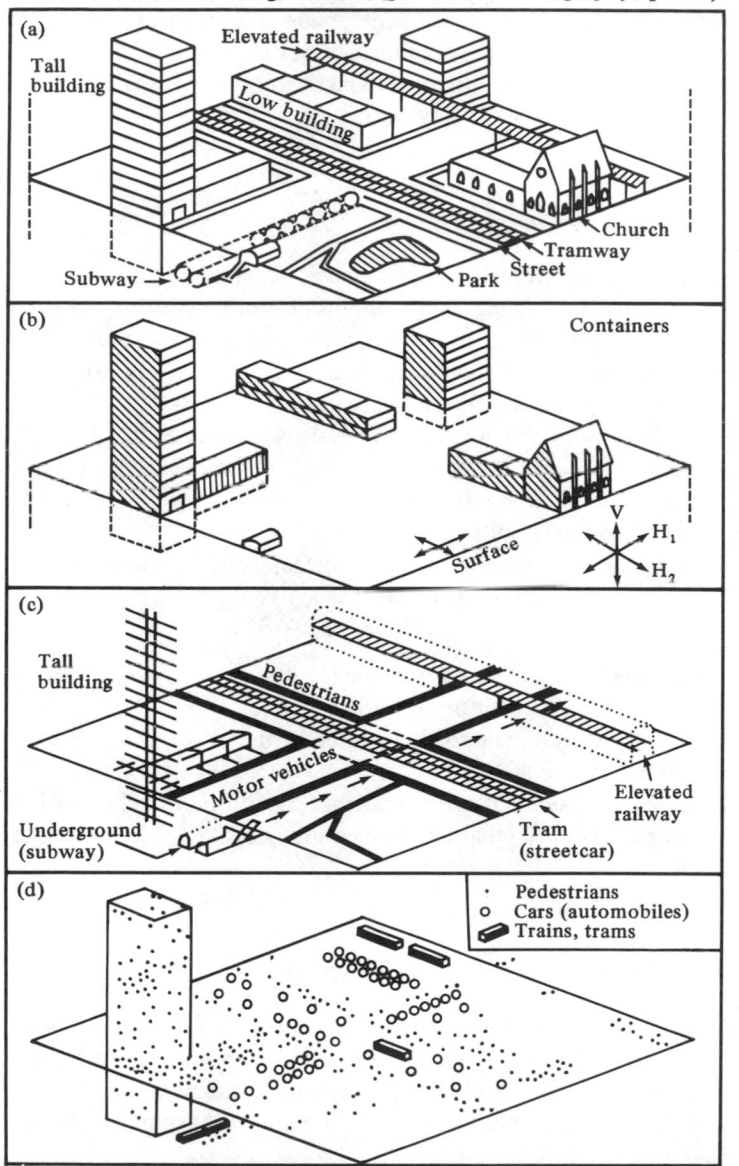

(5) *Time*

A fourth dimension—that of time—may be included at this stage imply-ing changes in location. This would include the movement of people and cars over a short period of time, and the much longer period changes in the distribution and form of all types of spatially-arranged phenomena. Such longer period changes are adjustments made as a result of changing socio-economic forces, which we examined in more detail in Chapter 1, where geographical systems were discussed. The short-term changes are an integral part of the forces at work in a system.

Correlation in geography

Correlation is very much at the centre of emphasis in the scientific geography. Fig. 1.1 highlights the importance of correlation between two (or more) variables which are felt to show significant co-variation, since it is from the results of such correlation analysis that generalizations may be drawn with a reasonable degree of certainty.

A full discussion of correlation analysis is given in S.I.G. 4, but it may be pointed out here that the graphing of a set of data such as that given in Figs. 1.2(a) and (b), where the data are plotted as x and y coordinates, will give three important results:

(a)   A precise description of the relationship between the two variables x and y, in the form of an equation such as

$$y = mx + c$$

where m is the gradient of the line (in the case of a straight line graph) and c is the intercept on the y axis.

(b)   A precise assessment of the degree of the relationship between the two variables x and y.

(c)   A precise statement of the statistical significance of the relationship, using probabilities that the set of readings could not have occurred by chance.

New views of distance: transformations

Fig. 1.1, box (5), shows where non-Euclidian views of space may be of some importance. When discussing networks, it is possible to ignore both distance and direction while the property of contiguity is analysed. Such manipulation is known as topology or 'rubber sheet geometry'. We may go much farther than this, however, as will be seen in the next chapter, for what is often most important in location is not what measured (Euclidean) distances really are, but what people *think* they are! Such perception of space may well be based on time distance or cost distance, or, indeed, upon some other aspect of economic activity such as income levels.

Further reading for Chapter 2

Cole, J.P., and King, C.A.M., *Quantitative Geography* (Wiley, 1968), especially Parts 1 and 2, and Chapter 13.

Haggett, P., *Locational Analysis in Human Geography* (Arnold, 1965), especially Chapter 3.

Haggett, P., *Geography: A Modern Synthesis* (Harper & Row, 1972), especially Chapter 14.

Kansky, K.J., *Structure of Transportation Networks,* Chicago Research Paper No.84 (University of Chicago, 1963)

Chapter 3

Transformations of space

We may consider geography as being mainly concerned with spatial distributions and their inter-relationships. In Chapter 2 we studied the role of mathematics in the scientific approach shown in Fig. 1.1 and saw that there are three main areas where mathematics is of use:

(a)   As a means of selecting information (sampling).

(b)   As a means of adding precision to the description of distributions.

(c)   As a means of describing correlation and indicating the significance of relationships.

If, when using the scientific approach, we find that no significant correlation exists (so that we cannot proceed to generalize), then it is possible that our original 'hunch' defining the problem was wrong. Such a situation would demand a complete rethinking of the problem.

On the other hand, if our hypothesis does not appear to stand up to testing, it could simply mean that the relationship between the two variables is obscured by *a basic misrepresentation of space itself.*

To make this clearer we can refer briefly to two models of distribution: the **fixed k system** of Walter Christaller (developed in the 1930s), and the very much earlier von Thünen model of rural land use, evolved from about 1806 onwards. In the first case, Christaller suggested that in those parts of the world not greatly touched by industrialization, villages, towns, and cities would be spaced evenly through the countryside. He then went on to say that each settlement would be one of a class of similar-sized settlements, each class providing a different order of goods and services, and the settlements in each class being evenly distributed throughout the whole area. In this way all the towns and villages could be placed in a hierarchy, with just a few really large towns, but with increasing numbers of settlements at each of the lower levels in the pyramid.[1]

In the case of the von Thünen model, rural land-use is supposedly arranged by type in concentric rings around urban centres. This pattern is related to the degree of competition for land, and to the relative ease with which the various products can be transported.[2] In both these models a first assumption was made that the earth's surface was everywhere the same, with no significant variations in resources such as fertility of soil and water availability. Such hypothetical uniform surfaces are usually referred to as being **isotropic**. The validity of such assumptions is often questioned, but such an approach does give some yardstick by which real distributions may be judged. However, the essential isotropic characteristics of both models may be summed up as:

(i) Each point in such landscapes has the same resource potential as every other. This refers to the possible variations, such as fertility, mentioned above.

(ii) Each point has the same degree of accessibility as every other point. In other words there are no physical or other factors which would bring about focussing of routes and therefore localized points possessing a high degree of accessibility.

This suggests that the variations in relief and other features which determine the degree of accessibility, and the variations in realized and capitalized potential which affect population density, may well have the effect of distorting true distances. Such true distances are a part of what we shall refer to as Euclidean space.

In later developments of both models some attempt is made to take into account variations in accessibility (easier transport along von Thünen's river),[2] and in potential (Losch's attempt at modifying Christaller's model in the light of industrialization, and therefore agglomeration).[3] These variations are seen as distortions of the original model. However, it is possible to work in reverse, as it were, and take the distribution in which we are interested, and try to iron out the distortions brought about by such factors as variations in accessibility and potential. If this proved successful we might then uncover otherwise hidden relationships which did not appear when we tried our original test.

Such straightening-out of distance is usually referred to as a **transformation**. William Bunge (1966) distinguishes between two types of transformation:[4]

(a) Morphological transformations (dealing with linear distances and accessibility).

(b) Density transformations (dealing with transformations of indicators of potential, such as population density).

It will be useful to examine these two types of transformation separately.

Morphological transformations

Many decisions concerning location (e.g. of factories, where to live, what crops to plant) are made in the light of a person's assessment of **distance**. In other words, consideration of distances from supplies of materials, from work place, from market, and so on, may prove of paramount importance. Distance in all these examples may be translated into terms of cost, of time, or of effort required to overcome the barriers set by distance. As we have seen, 'distance' is a basic geographic concept, and any movements have to overcome such barriers. The idea that distance presents a difficulty that has to be overcome in the working of a geographic system leads to the concept of **friction of distance**.

As already hinted, we quite naturally see distance in a number of ways.
'How far do you live from school?' 'About half a kilometre' (Euclidean distance).
'How far is it to Willesden Garage?' 'A 5p bus fare' (cost distance).
'How far is Victoria Station?' 'About 5 minutes walk' (time distance).

Even from earliest childhood we do not necessarily see distance in terms of yards, metres, miles, or kilometres, but quite often in terms of time or cost to overcome it.

Fig. 3.1 shows the fictitious town of Greenhurst, surrounded at various distances by 10 villages. The table below gives the road distance in kilometres from each village to Greenhurst (x column), and the percentage of each village's population that visited Greenhurst in the space of one month on shopping trips (y column).

Village	x distance from Greenhurst	Rank	y % pop. shopping in Greenhurst	Rank	difference in rank (d)	d^2
1	2·0	10	62	3	7	49
2	3·1	9	59	4	5	25
3	3·8	8	66	1	7	49
4	4·9	7	63	2	5	25
5	5·9	6	30	7½	1½	2¼
6	7·0	5	25	9	4	16
7	7·4	4	30	7½	3½	12¼
8	8·2	3	48	6	3	9
9	8·4	2	50	5	3	9
10	9·0	1	15	10	9	81

Either from examining these figures, or from general experience of such situations, we could hypothesize:

Fig. 3.1 *Greenhurst and its surrounding villages*

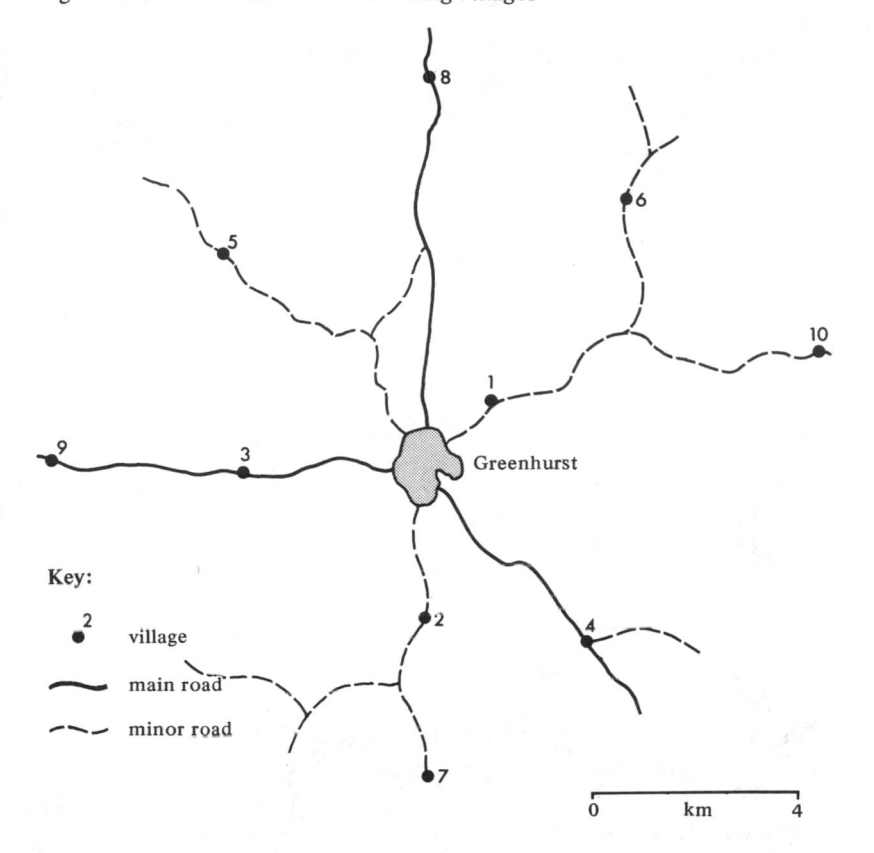

that the percentage of village population shopping in Greenhurst in the space of one month will vary inversely with the village's distance from the town.

In this way we have isolated two variables (distance and percentage of villagers shopping) in our study of shopping characteristics, and formulated a hypothesis.

If we now plot the data on a graph we can visualize the extent of the relationship between the two variables, and can then proceed to find the degree of correlation using regression techniques (S.I.G. 4, Chapter 7). The equation for the best-fit line, describing the relation mathematically, can also be calculated, as can the significance of the relationship.

To ascertain the degree of correlation, and its statistical significance, we could use, instead, Spearman's rank correlation coefficient, a much simpler method, also outlined in S.I.G. 4, Chapter 6.

Exercise 1 at the end of the chapter suggests that you work out the degree of correlation (using Spearman's rank correlation coefficient) between distance and percentage of population shopping in Greenhurst. The data should also be graphed to facilitate visual correlation. Fig. 3.2 gives the data in this form.

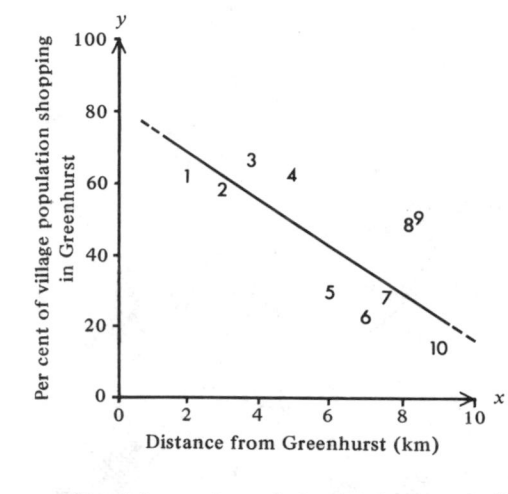

Fig. 3.2 *'Distance decay'; the percentage of villagers that visited Greenhurst in the space of one month. Each number represents one village in Fig. 3.1*

Visual inspection of the graph given in Fig. 3.2 casts doubt on the validity of our hypothesis. But despite the scatter of points on the graph, the Spearman coefficient (−0·69) does suggest that there is a significant correlation. (Tables to assess the significance of Spearman's coefficient are given in S.I.G. 4, Appendix 7. You may prefer to work through this exercise at a later stage when a full study of correlation and significance is made).

At this stage of the investigation we might feel dissatisfied that there are a number of surprising anomalies, villages 8 and 9 for example. Although we should be unlikely to discard our hypothesis completely, we might be interested in seeing whether we could 'straighten out' any possible distortions of space. Perhaps we feel that road distances are unrealistic by themselves and that we should take into account time of travel. In our example we could examine the type of roads serving the area, and perhaps allow for a speed of travel of 60 km/h for main roads, and 30 km/h for country lanes. Fig. 3.1 shows the Greenhurst area with its A roads and minor roads marked.

Having calculated the probable time of travel (allowing for, say, 10 minutes within Greenhurst for parking,etc.), we can arrive at the following set of data:

Village	x time-distance from Greenhurst	y % pop. shopping in Greenhurst
1	15	62
2	19	59
3	16	66
4	18	63
5	28	30
6	30	25
7	31	30
8	22	48
9	24	50
10	35	15

To what extent is this more valid? Exercise 2 at the end of this chapter directs you to work through to the significance testing stage using both graphical representation and Spearman's rank correlation coefficient.

This is a simple exercise introducing you to the idea of transformations of space in an attempt to reveal hidden relationships. If, in this case, the time transformation yielded no better results, the answer to the problem could lie elsewhere. Perhaps we should not have based our ideas on the movement of a substantially car-owning population. A bus timetable and route map might well give us the required information. This is where some research 'in the field' can clear up problems.

Using the map of the Greenhurst area we can also draw contours around the town. A piece of tracing paper placed over Fig. 3.1, with each village marked with its time-distance from Greenhurst could provide the base. The **isochrones**, or time contours (which are lines along which time from Greenhurst is constant), that we draw would be far from being concentric rings. The main roads, for instance, would mark out extensions of the time zones into the surrounding country.

An important question here is whether the isochrone map just drawn presents a true picture of the relationship of the villages to Greenhurst, in shopping terms at least. Perhaps a time-distance transformed map would present a closer approach to reality.

To do this it is suggested that you draw a series of rings centred on Greenhurst, with radii of, say, either 10 or 20mm, to represent time-

Fig. 3.3 *Details of mileages, directions from London, and times of fastest trains from London to selected towns in Britain (1968)*
(**Source**: Everson, J.A., and FitzGerald, B.P., *Settlement Patterns* p.125)

Town	Miles from London	Bearing in degrees from London	Fastest train time in minutes 1968
Aberdeen	523½	349	590
Aberystwyth	234¼	291	228
Barrow	264¾	325	304
Birmingham	110½	312	90
Brighton	50½	182	55
Bristol	118¼	270	110
Cardiff	145¼	270	132
Dover	77¼	111	90
Edinburgh	393	339	350
Glasgow	401½	325	405
Holyhead	263¾	304	275
Hull	196¾	356	202
Inverness	567¼	340	720
Ipswich	68¾	051	72
King's Lynn	97	013	124
Liverpool	193¾	318	155
Manchester	183½	328	150
Middlesbrough	238¾	348	224
Milford Haven	259½	275	381
Newcastle	268½	346	230
Norwich	115	036	120
Oban	502¾	329	716
Penzance	305¼	249	390
Plymouth	225¾	247	240
Southampton	79¼	234	70
Stranraer	451½	320	570
Wick	728¾	347	1160

distances of 10 minutes. Thus at a radius of either 40 or 80mm you would be at a distance of 40 minutes from Greenhurst, depending on the scale you had chosen. You can now put in each village at its correct time-distance and bearing from Greenhurst.

Fig. 3.4 *Time-distance transformed map of Britain using British Rail fastest train times from London as the basis. See Fig. 3.3 for data*

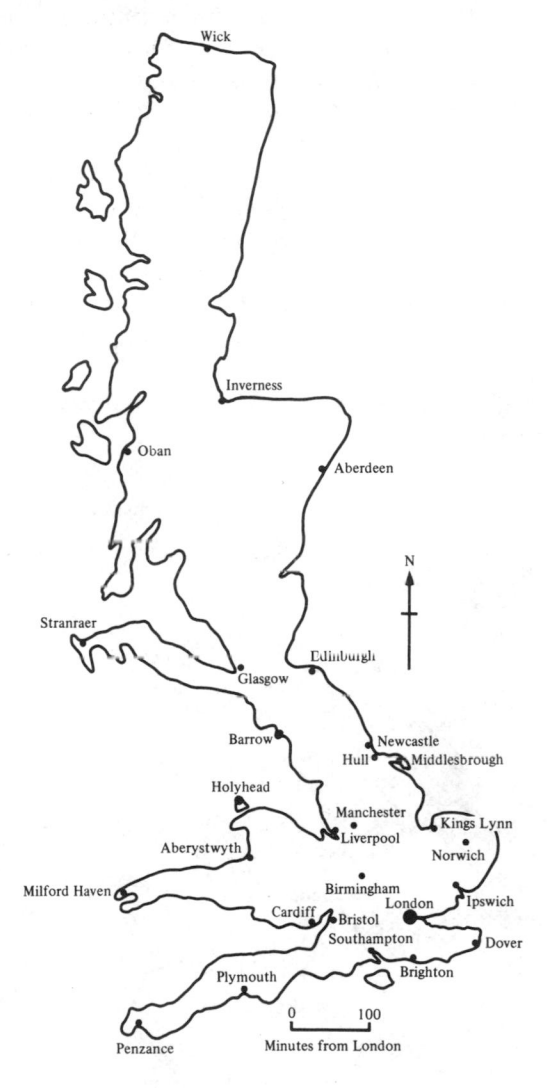

Such morphological transformations can bring to light relationships that otherwise may not be immediately apparent. Fig. 3.3, for example, shows British Rail journey times to a number of towns from London in 1968, together with their bearings from the capital. These figures are taken from

Fig. 79 in Everson and FitzGerald (1969), where the following idea is
more fully developed.[5]

If we take such timings as being indicative of general accessibility from
London (the London businessman's-eye view of Britain) we could perhaps
redraw the map of Britain with time-distance as our base. Such a redrawn
map is shown in Fig. 3.4. The principles involved in drawing such a map—if
you wish to try—are the same as for the Greenhurst transformation. It is
suggested you use a foolscap or A4-sized sheet of paper, and place
London close to the bottom right-hand corner. Rings, concentric
on London, may then be drawn at intervals of 10mm to represent
100 minutes rail travelling time, and then the towns plotted in at
their correct time-distances and bearings from London. The drawing of the
concentric rings is not, in fact, a necessity, so long as you preserve the

Fig. 3.5 *The development areas of Britain (Northern Ireland is excluded,*
but benefits from similar Government aid)

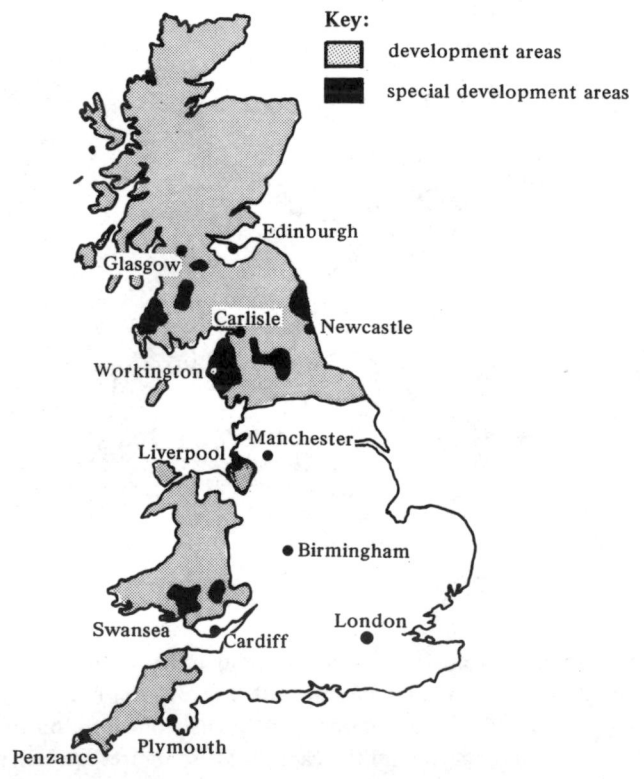

Fig. 3.6 *A part (adapted) of the Southern Region timetable from Gillingham and Maidstone East to Victoria*

Table 9

Weekdays

Miles from Victoria		BC							BC
36	**GILLINGHAM (Kent)**	07 29	07 37	..	..	..	07 46	07 54	08 12
34¼	**CHATHAM**	07 33	07 41	....	....	....	07 50	07 58	08 15
33¾	ROCHESTER	..	..	..	..	07 48	07 52	..	..
26¾	SOLE STREET	....	....	....	....	07 58	08 02	....	....
26	MEOPHAM	..	..	..	..	08 00	08 05	..	..
23¼	LONGFIELD	....	....	....	....	08 04	08 09	....	....
20½	FARNINGHAM ROAD	..	..	..	..	08 08	08 13	..	..
40	**MAIDSTONE EAST**	....	....	07 27	07 43	....	....	....	....
37½	BARMING	..	..	07 32	..	..	..	..	..
35¾	EAST MALLING	....	....	07 36	....	....	....	....	....
34¾	WEST MALLING	..	..	07 38	..	..	..	..	..
29¾	BOROUGH GN & W.	....	....	07 46	07 58	....	....	....	....
27	KEMSING	..	..	07 50	..	..	..	..	..
24	OTFORD	....	....	07 55	08 06	....	....	....	....
17¾	SWANLEY arr	..	..	08 08	..	08 13	08 18	..	..
	SWANLEY dep	....	....	08 08	....	08 14	08 18	....	....
14¾	ST MARY CRAY	..	..	08 13	..	08 19	..	..	..
10¾	BROMLEY SOUTH	....	08 10	08 20	....	08 26	08 28	....	....
—	**VICTORIA**	00 17	08 30	08H48	08 42	08 46	08H53	08 49	08 59

H – To Holborn Station, calls Blackfriars 3 mins earlier

BC – Buffet Car

correct time-distance scale of 10mm to represent 100 minutes. The last task, with the aid of an atlas, is to draw in the distorted coast line of Britain.

Compare the map you have drawn, or the one in Fig. 3.4, with the map of the Government Development Areas given in Fig. 3.5. To what extent does such a comparison reveal the problems of the old industrial areas and rural regions of western and northern Britain? The relatively economically depressed 'Grey' areas have their locations on the *edge* of Britain emphasized by this approach to accessibility. This differential shrinkage of Britain therefore highlights the problem of 'economic peripherality'.

Space transformations can produce peculiar quirks. Exercise 3 at the end of the chapter refers to Fig. 3.6, which shows a section of a timetable giving the times for London-bound morning rush-hour trains from the Medway area of Kent, and to Fig. 3.7, which is the corresponding map of this line to London. It is suggested that you carefully examine the time-distance map, which you are asked to draw, to see how a degree of **space-folding** or **inversion of space** has occurred.

The re-drawing of this North Kent Line is simple enough when we confine ourselves to a one-dimensional representation, i.e. a line. We run

Fig. 3.7 *The Southern Region lines from Victoria to the Medway towns.*
The number under the station name is the time (in minutes) taken to
London (Victoria) by the fastest available train to arrive between 8.0 a.m.
and 9.0 a.m. The fastest times from certain stations on the Maidstone East
Line involve changes at either Borough Green or Otford (see timetable given)

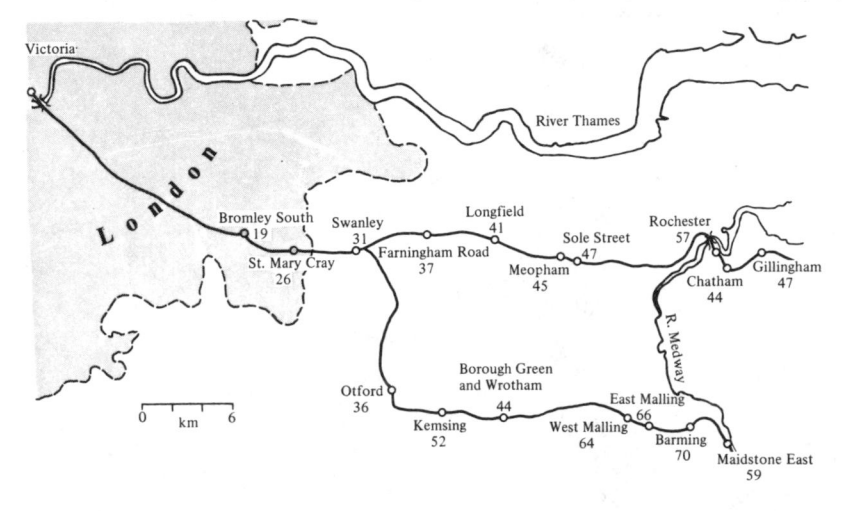

into great difficulty, however, if we attempt to draw a two-dimensional
map which includes inversions of space. This problem is being studied
currently by many cartographers interested in topology.

Two examples of transformations revealing hidden relationships may be
taken from the past. It has been suggested that the oceans and seas around
pre-industrial Britain may be transformed into an inland sea surrounded by
mainland Britain, across which most of British coastal and world trade took
place. The transformation (Fig. 3.8) may emphasize the importance of sea
routes at this time for overcoming friction of distance. Land routes were
very difficult, and those countries which had extensive seaboards had some
advantage over those which did not. The pre-eminence of coastal trade
(coal from Newcastle to London) comes out well.

The second example (Fig. 3.9(a)) is designed to show ancient Athens's
position in the littorally-arranged Greek world of 560 B.C. The importance
of the Mediterranean Sea is very strongly brought out by the pattern of
coastal settlements centred on Athens and Corinth. Fig. 3.9(b) gives
an Athenian's-eye view of the Greek colonies. In this instance space
has been transformed so that distance, instead of increasing linearly from
Athens, increases logarithmically, as shown in Fig. 3.10.

Fig. 3.8 *A transformation of the world map to emphasize the importance of 17th to 19th century sea routes*

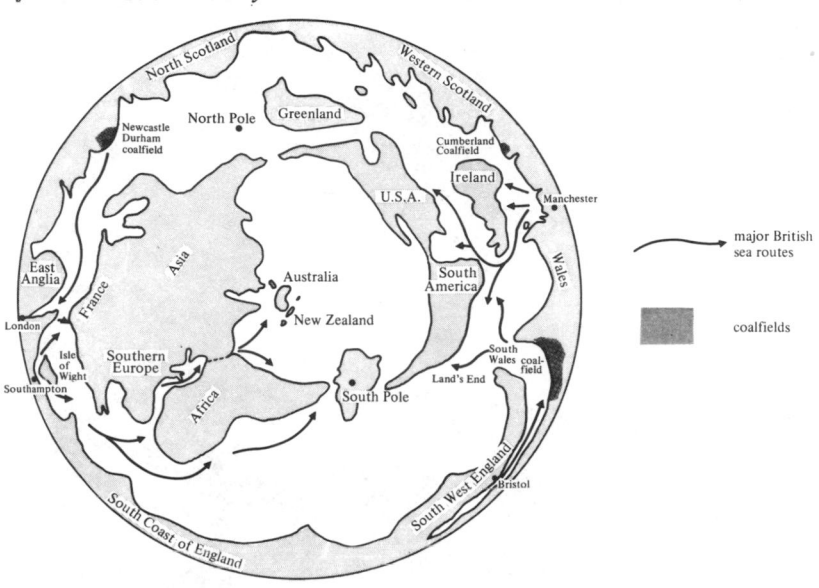

T. Hägerstrand, of the University of Lund in Sweden, was one of the first to use such a projection (Bunge, 1966, p. 54). He based his ideas on the apparent decrease at a logarithmic rate of transport costs with increasing distance. If this is true such a logarithmic transformation may give a more realistic picture of the spread of the Greek world.

The basis for such an assertion may best be described in terms of a hypothesis:

that it is twice as costly (or difficult) to travel 10km by boat as it is 1, 100km as it is 10, 1000km as it is 100, and so on.

If this is true, a map such as that in Fig. 3.9(b) may be justified because each unit of distance on the map increases the real world distance by a factor of ten. Thus to go from 1 to 100 on either of the axes is twice the distance from 1 to 10.

Density transformations

So far we have dealt only with distance transformations, in an attempt to bring some degree of order to the pattern of distributions and interactions on the earth's surface.

Fig. 3.9(a) *The Greek world in 560 B.C.*
(**Source:** McEvedy, C., *The Penguin Atlas of Ancient History* (Penguin, 1967), p.49)

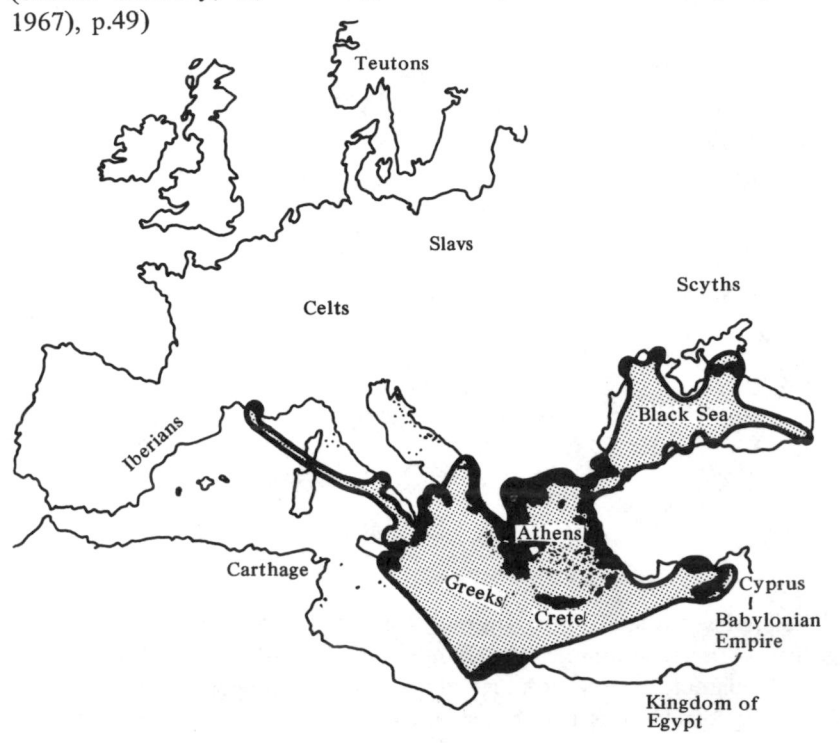

With density transformations the problem is not to untwist twisted space, but rather 'to spread uniformly the density of space' as Bunge expresses it. This approach assumes that there is not just Euclidean space, but also population space, income space, and so on. Any density map can, theoretically, be transformed. Fig. 3.11 shows a drawing of a three-dimensional model of the population map given in Fig. 3.12(a). In the case of a density transformation, the shape of the model suggests that the peaks should be flattened, to cause the areas of high density to take up a greater area, and the areas of low density to contract.

The result of such a transformation is shown in Fig. 3.12(a) and (b). Here a grid of squares has been placed over the map of the density of population in a rural area. The density surface is then 'flattened' so that the density of population is everywhere the same. This means that the 'surface' is now at a constant height, but the grid squares expand and contract

Fig. 3.9(b) *An Athenian's-eye view of the Greek world (560 B.C.). A logarithmic transformation centred on Athens. Distances from Athens given in miles. Greek area of influence shaded*

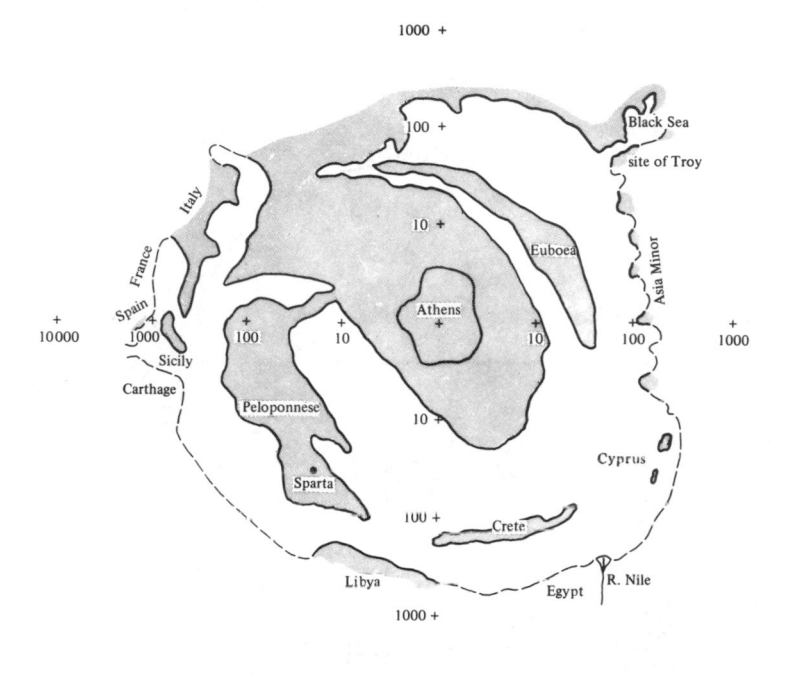

depending upon the number of people in each. The grid squares still contain the same number of people as they did before transformation, but they have become distorted so that, on the map, the density is everywhere the same. Thus Euclidean space is altered to allow 'population space' to be constant.

How may such a density transformation be used? Fig. 3.13 shows the distribution of post offices in the same rural area. Subjectively it could be described as being clustered, the degree of clustering being ascertained by further statistical analysis. But what principles underlie the distribution? If we assume that the post offices are all of the same size we might hypothesize that they will be related in some way to the distribution of population. Thus we would expect to find a regular, hexagonal distribution when the position of each post office is plotted on the population space map.

The process is to copy the transformed grid (Fig. 3.12(b)) and then

Fig. 3.10 *The logarithmically transformed graticule on which Fig. 3.9(b)*
is based

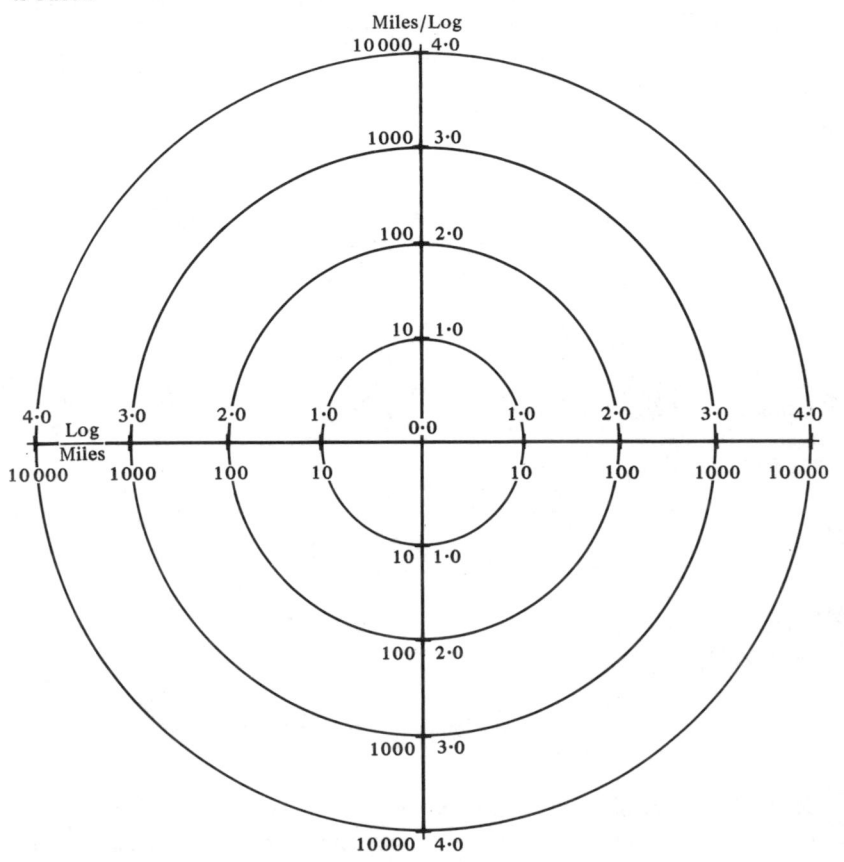

mark in the location of each post office with respect to the pattern of grid
lines.

Exercise 4 at the end of the chapter suggests that you attempt to
relocate the post offices on the transformed grid (Fig. 3.12(b)) and to
comment upon the pattern that is revealed. Fig. 3.14 shows the transformed
pattern.

This example of a density transformation is a fictitious one, obvious in
character and given simply to illustrate the principles involved. Professor
Arthur Getis explains in an article (Ambrose, P. (ed.), 1969) how he tested
his ideas in the southern part of the city of Tacoma in Washington, U.S.A.
In this study Getis related the distribution of supermarkets not just to

Fig. 3.11 *Population density surface in a fictitious area. x, y, and z represent population densities*

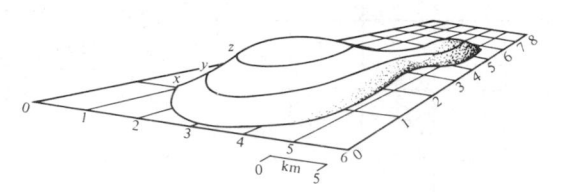

Fig. 3.12 *Transformation of population density map to make population space constant*

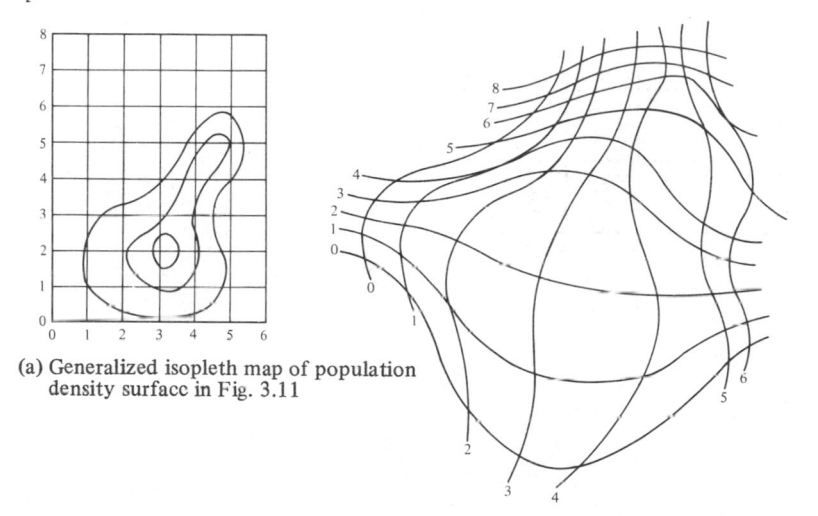

(a) Generalized isopleth map of population density surface in Fig. 3.11

(b) Grid squares transformed according to population density

population space, but also to 'income space'. The problem was 'to determine the effect of the distribution of consumer disposable income for retail goods on the location of retail stores'. To do this Getis had to make use of census material for each of the census tracts within the study area, to work out, not only the population total, but also the amount of income available for groceries. This figure was calculated by working out for each census tract the proportion of the population that fell within each income group. This then enabled Getis to work out how much would be available for spending on groceries.

The next stage was to plot the calculated 'consumption expenditures available for groceries' on to a gridded map as shown in Fig. 3.15. Here

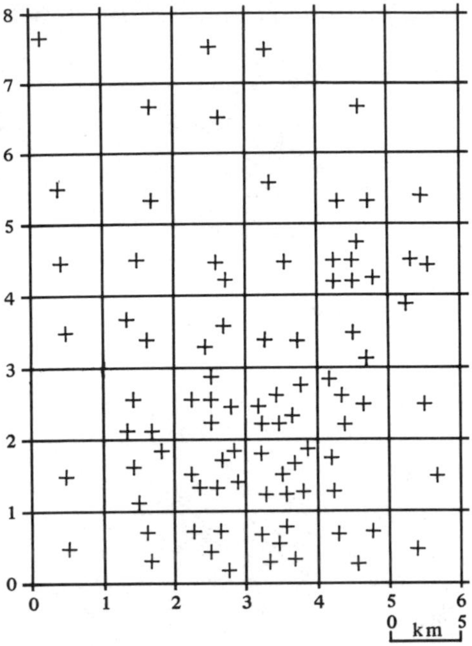

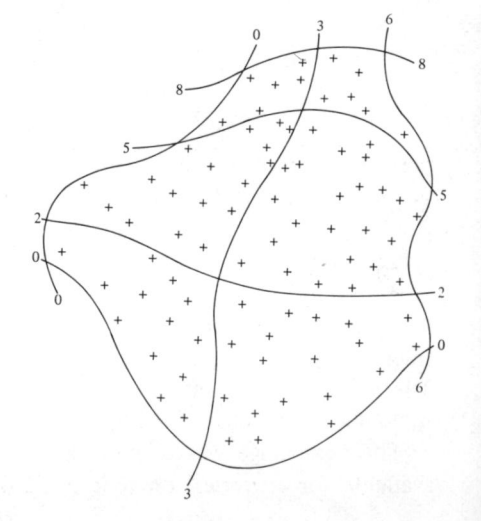

Fig. 3.13 *Distribution of post offices in the fictitious rural areas shown in Figs. 3.11 and 3.12*

Fig. 3.14 *Distribution of post offices (see Fig. 3.13) transformed according to population space*

Fig. 3.15 *Location of consumption expenditures available for groceries in a part of Tacoma, Washington, U.S.A., in thousands of dollars.*
(**Source**: Getis, A., 'The determination of the location of retail activities with the use of a map transformation' in *Economic Geography*, 39 (1963), pp.14–22)

	10	11	12	13	14	15
10	235	321	611	449	462	496
11	543	663	764	472	612	400
12	457	827	940	535	264	1023
13	345	827	823	420	288	162
14	194	515	680	265	298	46
15	24	397	508	336	153	99
16	35	312	300	31		
17	43	251	263	169		
18	87	147	175	88		

each grid square contains a figure representing the estimated expenditure (in thousands of dollars) available for groceries for that square.

The grid cells were then transformed, as Getis says, 'so that a unit of area anywhere on the distorted map would be equal to the estimated consumption expenditures for groceries' as shown in Fig. 3.16. 'Thus we note' he goes on to say 'that cell 10-10 with consumption expenditures for groceries of $235,000 is one half the size of cell 10-14 with $462,000 available for groceries'. In this way the consumption expenditures are spread evenly throughout the study area by distorting Euclidean space.

Getis then puts forward the following rules which he used in producing the map in Fig. 3.16:
(a) Cells should be rectangular whenever possible.
(b) All cells which are contiguous on the original map must also be contiguous on the distorted map.
(c) The size and shape of the distorted map must be the same size and shape as the original map.

Rules (a) and (c) were not observed in transforming Fig. 3.12(a) because the total number of grid squares was small, and problems of contiguity at the periphery were therefore not great, and because the form and nature of the transformation were more readily apparent.

Getis then took the transformed base map shown in Fig. 3.16 and plotted on this the location of 12 supermarkets evenly spaced (at the centres of hexagonal market areas, which he was hypothesizing should exist). They

Fig. 3.16 *Map distortion of consumption expenditure for groceries in Tacoma study area* (**Source**: as Fig. 3.15)

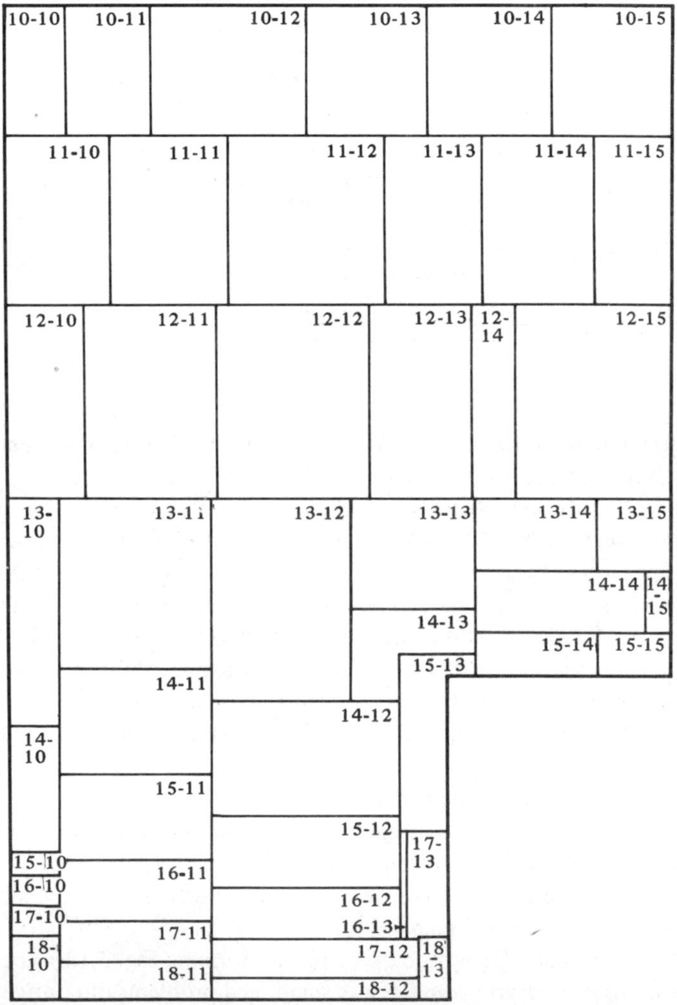

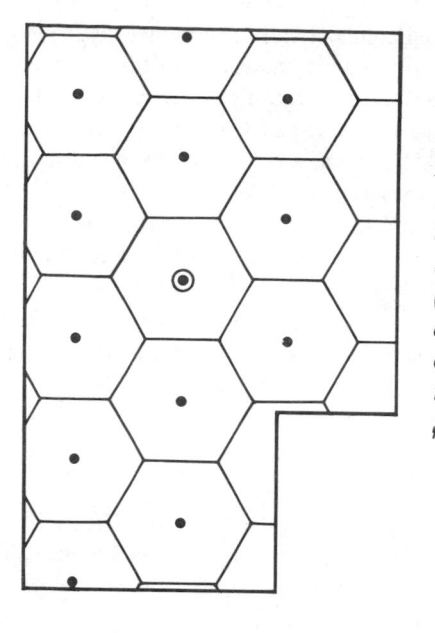

Fig. 3.17 *Theoretical hexagonal trade areas placed on transformed base (shown in Fig. 3.16)* (**Source**: as for Fig. 3.15) *N.B. 12 supermarkets were chosen (11 areas plus two halves) as the average annual sales of a supermarket were then (1963) a little over $1 500 000, and computed consumption expenditures for groceries in this area were approximately $18 000 000*

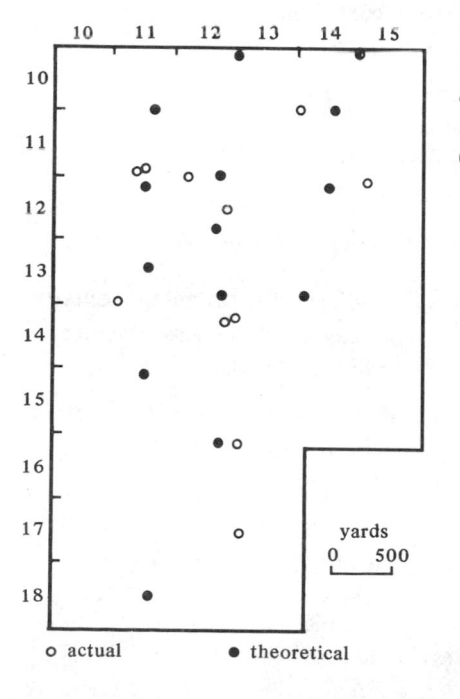

Fig. 3.18 *Theoretical and actual locations of supermarkets in study area* (**Source**: as for Fig. 3.15)

o actual   • theoretical

were then transformed to their predicted positions (respective to the grid lines) on the undistorted grid (see the change in position of the black dots between Fig. 3.17 and 3.18). The actual locations of the supermarkets (open circles) were then added (see Fig. 3.18) so that appropriate correlations could be made. The result, at this stage, was that the average distance between predicted and actual locations was under 400 metres.

Getis adds further refinements, and then states some of the limitations of the method, but his article is primarily dealing with technique rather than with subject matter. Nevertheless a more careful inspection of the article would help to point out the way in which transformations of space may be of help in investigating geographical relationships. The article is easily available in reprinted form in Ambrose (1969), pp. 204–15.

Exercises

1. (a)   Plot the data given on p. 44, concerning distance from Greenhurst and percentage of village population shopping in Greenhurst, on a graph. Describe the form and degree of the relationship you observe.

   (b)   Use Spearman's rank correlation coefficient,

$$r_s = 1 - \frac{6 \cdot \Sigma d^2}{n(n^2 - n)}$$

   where d is the difference in rank,
         n is the number of villages (10),
         Σ is the instruction to sum,

   to work out the degree of correlation, and whether it is positive or negative.

   Reference to S.I.G. 4, Chapters 3 and 6, will give further details concerning correlation and significance, and you may prefer to leave these exercises until the relevant sections of the series have been studied.

2. Analyse the data given on p. 47 for the **time**-distances of villages from Greenhurst in the same manner as in Exercise 1, above.

3. Examine Fig. 3.6 (p. 51) which shows the timings of London-bound morning rush-hour trains from the Medway area of Kent, and Fig. 3.7, which shows the map of the line into London with stations at their true distance apart.

   (a) Re-draw the map of the line using time-distance as a basis. A scale of 20mm to represent 10 minutes should prove suitable.

(b) Comment on the change in pattern observed.

(c) In what ways could this South-Eastern commuters'-eye view of London affect the pattern of housing development in the region?

4. (a) Trace the grid given in Fig. 3.12(b) onto a piece of paper.

(b) Carefully re-locate the position of the post offices given in Fig. 3.13 onto the new transformed grid.

(c) (i) Describe the difference in pattern of post offices between Fig. 3.13 (untransformed) and the transformed map you have produced from question 4(b).

(ii) What geographical explanation can you give for this fundamental change in distribution?

(Note that Fig. 3.14 gives the transformed pattern.)

References

1. Christaller, W., *Central Places in Southern Germany*, translated by C.W. Baskin (Prentice-Hall, 1966).

   See also:

   Garner, B., 'Models of Urban Geography and Settlement Location', in Chorley, R.J., and Haggett, P. (eds.), *Socio-Economic Models in Geography* (Methuen, 1968).

   Everson, J.A., and FitzGerald, B.P., *Settlement Patterns* (Longman, 1969), Chapter 9.

2, Hall, P. (ed.), *Von Thünen's Isolated State* (Pergamon, 1966).

   See also:

   Chisholm, M., *Rural Settlement and Land-Use* (Hutchinson, 1962), Chapter 2.

3. Garner, B., 'Models of Urban Geography and Settlement Location', in Chorley, R.J., and Haggett, P. (eds.), *Socio-Economic Models in Geography* (Methuen, 1968), pp.315—19.

4. Bunge, W., *Theoretical Geography*, Lund Studies in Geography (Gleerup, 1966), p.270.

5. Everson, J.A., and FitzGerald, B.P., *Settlement Patterns* (Longman, 1969), Chapter 11.

Further reading for Chapter 3

Abler, R., Adams, J.S., and Gould, P., *Spatial Organization* (Prentice-Hall, 1971), Chapter 3, pp.72—88.

Ambrose, P. (ed.), *Analytical Human Geography* (Longman, 1969), Section 5, particularly: Thompson, D.L., 'New concept: subjective distance', reprinted from *Journal of Retailing*, 39 (Spring, 1963), and Getis, A., 'The determination of the location of retail activities with the use of a map transformation', reprinted from *Economic Geography*, 39 (1963), pp.14–22.

Berry, B.J.L., *et al.* (eds.) *Spatial Analysis* (Prentice-Hall, 1968), particularly Tobler, Waldo R., 'Geographic Area and Map Projections', pp.78–90, reprinted from *Geographical Review*, 53 (1963).

Bunge, W., *Theoretical Geography*, Lund Studies in Geography (Gleerup, 1966), especially Chapters 6 and 7.

Everson, J.A., and FitzGerald, B.P., *Settlement Patterns* (Longman, 1969), especially Chapter 11.

Haggett, P., *Geography: a Modern Synthesis* (Harper and Row, 1972), especially Chapter 4.

Haggett, P., *Locational Analysis in Human Geography* (Arnold, 1965), especially pp.38, 53–5.

Chapter 4

Perception and decision-making in geography

The perceived environment

In Chapter 3 we looked at the distortions of space which can be measured (e.g. the exercise producing the rail-time transformed map in Fig. 3.4), or estimated (e.g. the exercise leading to the construction of the Athenian's-eye view of the Greek colonization given in Fig. 3.9(b)). In these instances the transformed maps represent a conscious or sub-conscious view of the surrounding world. Some such views of the surrounding world are often very much at odds with what is actually there. These views are of the world as it is **perceived** by people, and it is *this* world environment which affects decision-making. In other words, people make geographical (or any other) decisions in the light of what is *perceived* rather than what actually *is*. At a very simple level one could say that a decision whether or not to locate permanent settlement near a river is based not on the fact of flooding, but rather on man's awareness or perception of the likelihood of flooding.

An excellent overall discussion of the way in which the environment is perceived is given by H.C. Brookfield in his article 'On the environment as perceived' (Board *et al* (eds.), 1969).

Brookfield makes the point that man perceives his environment and thereby receives a stock of information related to his needs. This information then forms the basis for decision-making. He goes on to say, 'Activities initiated as a result of decision yield results; these are further evaluated to modify future decision, both directly and by modifying the perceived environment and hence the perceived resources.'[1] Such a situation as this is neatly related to an expression of man's activities in terms of the systems analysis already referred to (pp.17–23). The perceived environment (as opposed to the so-called 'real' environment) may be regarded as a sub-system operating in the existing 'real' world as a result of the decisions which are made. Continuing the theme of a systems approach, Brookfield points out that:

'If there is no change in the real environment through time, no change in means available, no new imported or locally generated information, and constant population number, the sub-system would in time achieve a steady state, or even develop some of the characteristics of a closed system, including entropy, as perceived environment and perceived resources ceased to undergo futher modification.'[1]

The latter tendency is unlikely to occur in the world today where communication is both rapid and sophisticated, and where technology is rapidly advancing, but as he suggests, situations close to this must have been prevalent during the Palaeolithic era, and existed until quite recently in certain isolated areas populated by culturally primitive groups of people; a form of entropy would exist where there was no change or development in the culture and life-style of the people concerned.

Brookfield also discusses the role of **resources** in the perceived environment. He suggests that they can, of course, be regarded as a fact of the real environment, but that there may be distinct advantages in regarding them as an evaluation placed on the perceived environment. He suggests that 'The resources of a place at a given date are an evaluation of the perceived environment as of that date, which in turn is a model of the real environment produced by using the stock of information available on that environment.'[2] This quotation is important in that it strongly implies that the perception of a resource may vary over a period of time. This is despite the fact that the resource itself does not alter. Alaska, for example, has been viewed from a resource point of view in many different ways: as being virtually of no use by the Russians prior to its sale to the United States in 1867 for $4.14 per square kilometre; as a wealthy source of gold exploitation in the late nineteenth century; as a strategic frontier by the U.S. Pentagon in the 1950s and 1960s; and more recently as a rich source of oil.

In some instances, of course, the resource was hidden (the oil in Alaska) until some comparatively late date, but in many cases the resource has been apparent and available but not exploited until recently, due to such factors as insufficient technological skill or unsympathetic economic environment. In other words man's awareness of his scientific and economic environment had not evolved far enough for him to perceive those parts of his physical environment as useful resources.

The perception of hazards

A particular area in which decision-making in the light of perception of environment has been studied is the evaluation of **hazards**. In examining the kind of decision concerning, for example, whether or not to build a permanent settlement near a river, we must think in **probabilistic** terms. That is, we can only assign certain degrees of probability to the outcomes of man's decisions. Some may venture risks, others will be more reticent or conservative in their choices. Nevertheless if we examine the behaviour of many individuals we may find a considerable degree of regularity, although the results of such decision-making may show quite a wide diversity. This degree of regularity is what is being sought in Fig. 1.1 (box 4) on page 2.

Several geographers have seen such decision-making (in a hazardous context) as a game, perhaps a deadly earnest game, against the environment. As a result, a branch of mathematics known as **game theory** is being applied more and more to the decision-making behaviour of individuals. In many parts of the world extensions of farming into areas of considerable risk have been made. The pioneer wheat farmers in North America had to make decisions in such a situation or 'game' where they were in almost total ignorance of the complete range of vindictiveness of the environment.

Even in simple subsistence economies the choice of which crop combinations to plant, in terms of vicissitudes of climate, presents many problems. Where hazards are great the peasant farmers may well choose a rather conservative game **strategy**—they will tend to plant those combinations of crops which appear most likely to stave off hunger. Such a **satisficer** approach is a defensive strategy and is unlikely in the long run to produce large returns. This contrasts strongly with the Western farmer farming for profit; he has a more efficient **information field** and is thus able to attempt to maximize his returns. This is the **optimizer** approach.

An early work on this theme was P.R. Gould's article 'Man Against His Environment: A Game Theoretic Framework', easily available in reprinted form (Ambrose, 1969, pp. 243—57). Abler, Adams, and Gould discuss the use of a games approach to geography in an excellent summary in their book (1971, particularly pp. 478—89).

Flood, drought, and earthquake are amongst the most dangerous of hazards faced by man. A pioneer geographical work was that by T.F. Saarinen, *Perception of Drought Hazard on the Great Plains* (1966). In a case study of six counties on the Great Plains in a transitional zone between the humid areas of the east and the drier areas of the west, Saarinen found that all the farmers under-estimated the actual proportion of drought years. As Saarinen says:

'They (the wheat farmers) tend to under-estimate the frequency of drought years, and to be over-optimistic about the number of very good years and the size of crops in such years. All but the most recent, the most severe, and longest droughts tend to be forgotten, though individuals may recall the first one they experienced.'[3]

Abler, Adams, and Gould discuss Saarinen's work in some detail and comment that:

'Much of the history of wheat farming in the Great Plains and other marginal areas of the world settled by Europeans can be written in terms of over-optimism and degrees of risk-taking that approach wild-eyed speculation. Some pioneer farmers appear willing to probe to the outermost limits of a marginal environment, perceiving the opportunities through rose-tinted

Fig. 4.1 *Bournemouth mental map*

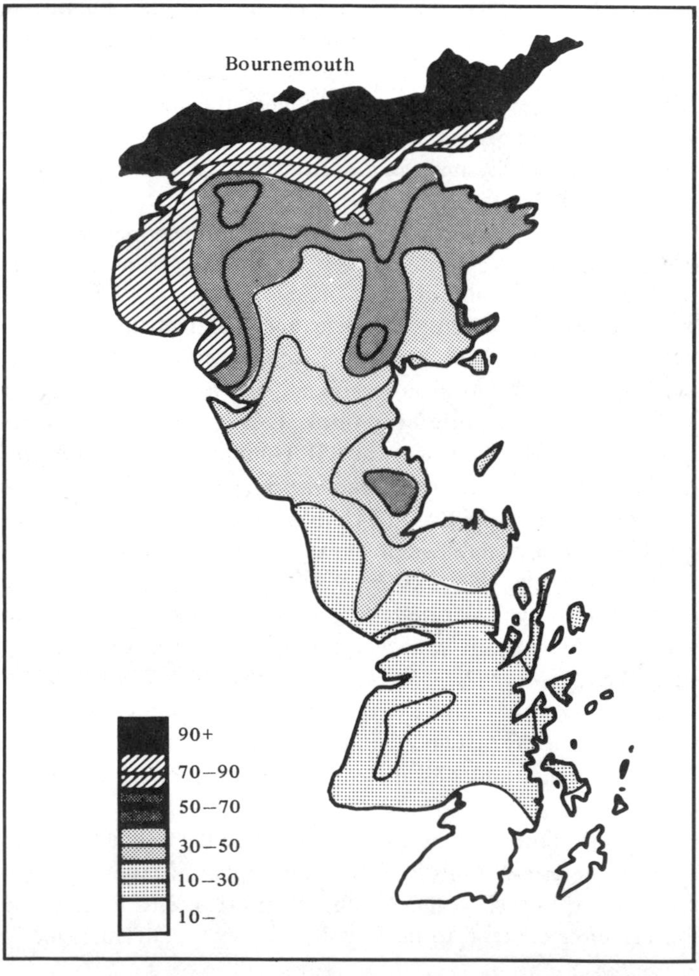

spectacles until an arid year sends them scurrying back to safer areas or forces them to hang on grimly waiting for better times to come. In the wheat areas of Canada and southern Australia the wheat lands have "pulsed" not only in response to transportation development and fluctu-

Fig. 4.2 *Inverness mental map*
(**Source**: Figs. 4.1, 4.2, and 4.3 are drawn after maps illustrating an article by Rodney White, 'Mental Maps of Britain', *New Society*, 327 (2 January 1969). Data for the maps were provided by Peter Haggett)

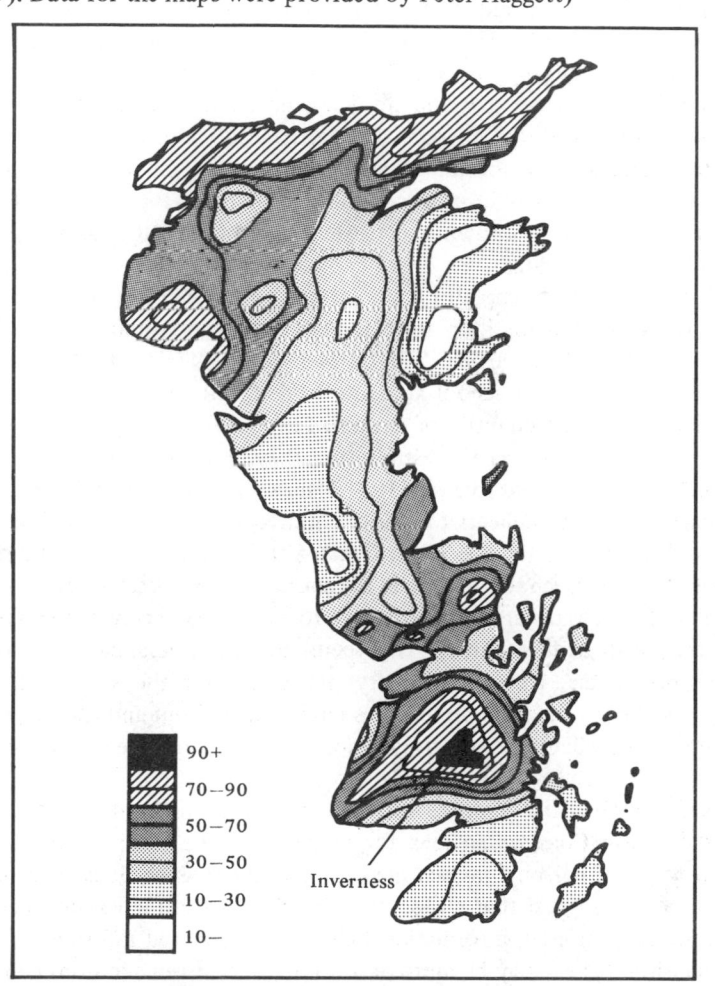

ations of world prices, but also according to the perception of men, whose memory traces of environmental hazard fade with time only to be reinforced by a crippling year of drought and dust storms.'[4]

Migration decisions and the perceived environment

The result of the limitation of one's perception of the environment or field of information is illustrated by family migration habits. In a recent school survey it was found that over a period of 20 years approximately 75 per cent of the household moves were made within a distance of 5 kilometres.[5] This meant that most moves were made within a locality known to the family. The sample taken was very small (70 families) but the results did suggest that such location decisions are more usually made within the context of that which is known, rather than that which is unknown. In this case the exercise was partly invalidated as no attempt was made to assess the anchoring effect of not changing job, or not wishing to make a move that would upset a pupil's schooling. Nevertheless it is fairly certain that what is known and trusted is an important consideration in such location decisions.

An interesting example of a similar perception study with, perhaps, an important bearing on how people migrate within Britain was given by Rodney White in an article 'Mental Maps of Britain'[6]. He described an experiment in which school students were asked to rank in order their preference for the counties of Britain. They had to place the counties in order from 1 to 92 on the assumption they could have a free choice of where they would like to live. Fig. 4.1 shows how the mental desirability of parts of Britain appears to school children in Bournemouth. There is a marked preference for the South Coast. On the other hand Fig. 4.2 shows the preference of school children in the county of Inverness. Again a marked local preference appears, but there tends to be a strong secondary preference for the south of the country. This seems to be a general pattern; a strong local preference, with a secondary preference for the south. Thus, as Fig. 4.3 shows, the local preferences tend to cancel out and leave a general desirability gradient from south to north, with a slight peak for the Lake District and a hollow 'sink' for London.

Remembering the continued drift to the South-East, and the popularity of the South Coast as an area for retirement, one can perhaps see the importance of personal mental maps in decision-making. It is then only a small step to suggest that all location decisions are made within a certain, necessarily restricted, information field. Such perceived information used by decision-makers may be more or less imperfect. Even decisions concern-

Fig. 4.3 *National mental map*

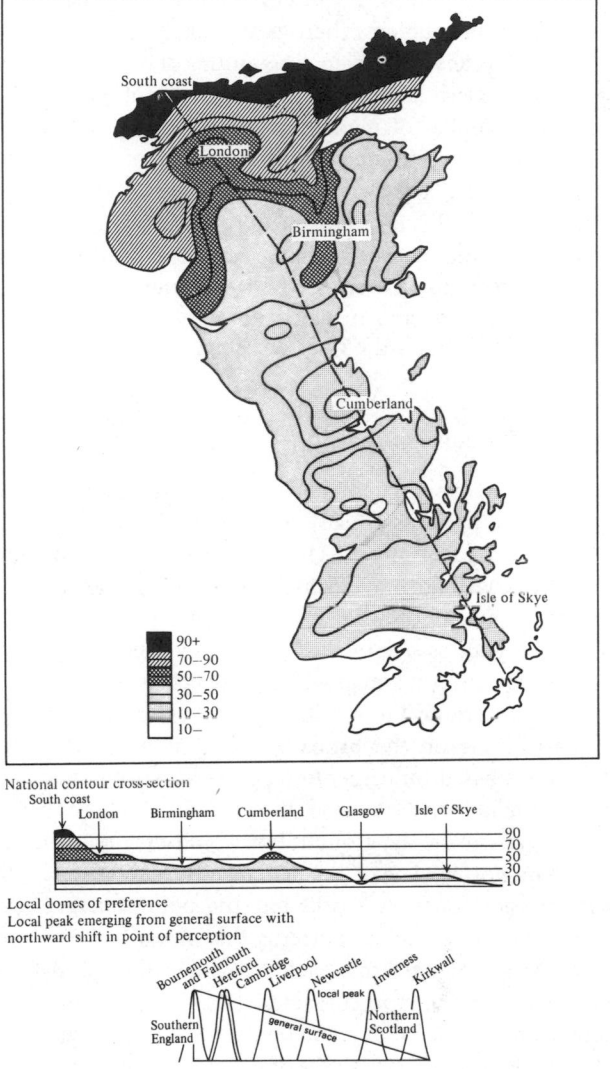

ing the location of modern industrial plants are made in the light of less than perfect knowledge of all economic, social, and physical conditions, despite the advice of economists, accountants, and market research workers. The passage of time too, brings an increasing uncertainty into such prediction.

This brings us to another important point about the decision-making behind geographical location. Even with as much information as possible concerning a location problem, there is still likely to be an infinitely large range of reasonable answers, even discounting those thought undesirable from economic or social standpoints. The essence of good planning is the preparation of a number of plans, each with its pros and cons. Decision-making therefore involves discussion of a number of possibilities, and the outcome may well be swayed by the influence of pressure groups and vested interest, even if this is not an entirely conscious element.

The fact remains that decision-making concerning location is very rarely concerned with one right answer. A whole spectrum of answers is possible, many better than others, and it may be a matter of chance persuasion or personal whim which guides the final decision.

Simulation and decision-making

The points made above concern current activity, but the same approach may be applied to patterns already established. Take the example of the road pattern on a part of any $1''$ Ordnance Survey map. Of course we can recognize routes twisting to avoid hills, focussing at the ends of estuaries, following the alignment of what appear to be river terraces, and so on.

Suppose that you, as a group of geography students, were given just those physical elements on a map which appear to affect the route pattern. Assume that you have no knowledge of the sheet under discussion. If you were asked to decide on the probable route network, would the several different versions based on perception of the map environment by individuals in the class be unreal? Of course, the patterns would probably all differ from the one present on the ground, but would they not be as valid as the real one? Perhaps not, in view of the lack of information concerning factors other than physical ones at work, but the point emerges that the real pattern is just one of countless patterns that *could* have evolved. It is just one of a whole set of patterns, each of which, though different, would preserve a number of common characteristics.

A reference to the idea that spatial patterns may be created 'artificially' leads to a consideration of the study of geography by means of **simulation**. Simulation techniques, where the development of geographical patterns is studied by setting up an artificial situation to parallel real spatial development through time, are of use in revealing the way in which the patterns have actually developed, and thereby, perhaps, allowing predictions of future change and progress.

A route and town location simulation is given in Everson and FitzGerald (1972), p.17. In the exercise a map of a fictitious country called Urbania shows rivers, mountains and hills, and villages. You are asked to mark in the expected road system and the location of the capital, ports, and market towns. It is a very simple simulation in which a comparison of 15 or so unique patterns produced by a group would reveal an underlying similarity. This would reflect equifinal behaviour in the systems at work.

Although 'chance' is not actually built into this simulation as a specific factor, many chance factors do operate and 15 different perceptions of the situation appear as 15 unique solutions. Other types of simulation build in 'chance factors' (referred to on p.76), which are *seen* to operate significantly, or may use chance (or **stochastic**) principles to develop the patterns being investigated.

An example of such a stochastic technique is the use of the **Monte Carlo method** to simulate the spread of such things as ideas, urban growth, and settlement in new lands. The simulation starts with the assumption that, for example, a farmer has knowledge of some new technique which he is willing to pass on to his neighbours. There is, of course, a greater chance of his 'telling' neighbours who are close, than those who are far away. This can be defined as a **communication field**, perhaps more commonly called an **information field**, outside which the probability of a telling is so low as to be negligible.

Such a communication field can be divided into a grid of squares or hexagons, in which each square (or hexagon) can be assigned a number representing its probability of receiving a telling. A **probability matrix** is shown in Fig. 4.4(a). All the squares should total 1·00, this representing the absolute certainty of a telling somewhere on the matrix. Squares close to the centre of the matrix would have a high probability, squares farther away a lower probability.

This probability matrix can be converted from one with a statement of probability to one where a sequence of numbers from 000 to 999 has been substituted for the probabilities. Fig. 4.4(b) shows such a conversion corresponding to the matrix in Fig. 4.4(a). A probability of 0·010 in Fig. 4.4(a) represents 10 chances in 1000, therefore 10 three-digit numbers, for example, 648—657 would be placed in the corresponding square. The first square should ideally start at 000 and the final square on the matrix finish at 999.

The completed matrix (Fig. 4.4(b)) is then placed on the map centred on the location of the first teller and numbers are then drawn from a

Fig. 4.4(a) *Probability matrix; in each square is a statement of the probability of a hit or a 'telling' which originated from* **X** *in that square*

				·001	·003	·003	·001				
		·001	·005	·008	·010	·010	·008	·005	·001		
	·001	·008	·010	·014	·020	·020	·014	·010	·008	·001	
	·005	·010	·021	·018	·013	·013	·018	·021	·010	·005	
·001	·008	·014	·018	·005	·005	·005	·005	·018	·014	·008	·001
·003	·010	·020	·013	·005	**X**		·005	·013	·020	·010	·003
·003	·010	·020	·013	·005	**X**		·005	·013	·020	·010	·003
·001	·008	·014	·018	·005	·005	·005	·005	·018	·014	·008	·001
	·005	·010	·021	·018	·013	·013	·018	·021	·010	·005	
	·001	·008	·010	·014	·020	·020	·014	·010	·008	·001	
		·001	·005	·008	·010	·010	·008	·005	·001		
				·001	·003	·003	·001				

random number table, three digits at a time. The final digits of numbers in a telephone directory may be used instead of the tables. Each three-digit number locates the telling on the matrix, and therefore on the underlying map.

There are now two potential tellers on the map, and the probability matrix is placed over each in turn to generate two more potential tellers which represent a second **generation**. The four now present can each give rise to another, the four new ones being the third generation. Eight tellings are now present and the process may continue for as many generations as required. Such a pattern developed to the fifth generation is given in Fig. 4.5. Various barriers to such diffusion may be produced by limiting the proportion of tellings that can have effect in certain areas.

Fig. 4.4(b) *Probability matrix for generating a diffusing pattern such as the spread of settlement* (**Source**: *Everson and FitzGerald* (1969), p.14)

				000	001–003	004–006	007				
			008	009–013	014–021	022–031	032–041	042–049	050–054	055	
		056	057–064	065–074	075–088	089–108	109–128	129–142	143–152	153–160	161
	162–166	167–176	177–197	198–215	216–228	229–241	242–259	260–280	281–290	291–295	
296	297–304	305–218	319–336	337–341	342–346	347–351	352–356	357–374	375–388	389–396	397
398–400	401–410	411–430	431–443	444–448	**X**		449–453	454–466	467–486	487–496	497–499
500–502	503–512	573–432	533–545	546–550			557–555	556–568	569–588	589–598	599–601
602	603–610	611–624	625–642	643–647	648–652	653–657	658–662	663–680	681–694	695–702	703
	704–708	709–718	719–739	740–757	758–770	771–783	784–801	802–822	823–832	833–837	
		838	839–846	847–856	857–870	871–890	891–910	911–924	925–934	935–942	943
			944	945–949	950–957	958–967	968–977	978–985	986–990	991	
				992	993–995	996–998	999				

Note that the numbers between 0 and 9 should be selected only when preceded by two noughts. Likewise, numbers between 10 and 99 should only be selected if preceded by one nought. In other words *all* numbers should be selected as three digits from the random number table

This would be similar to new ideas falling on deaf ears, the spread of foot and mouth disease into a dominantly arable region, or the migration of people through a mountain barrier with well-defined gaps. The Monte Carlo technique is especially useful in simulating **spatial diffusion** (which can, perhaps, be regarded as one of geography's basic concepts) and lends itself well to operation using a computer, which can speed up the process immensely. This computer use of the technique could have important repercussions for planning.

Any technique using chance as an important factor is referred to as being stochastic. The first geographer to make use of stochastic techniques

Fig. 4.5 *Simulated diffusion (e.g. spread of settlement or of foot and mouth disease) from an origin at ⊗ , developed to the fifth generation)*

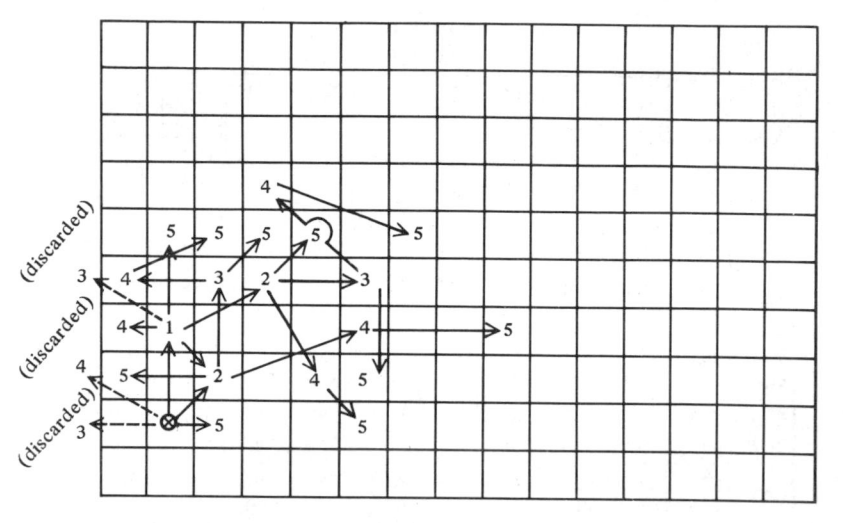

was T. Hägerstrand who, in the 1950s, studied the effects of Swedish farming innovations in the late 1920s. This process of predicting the course of past events is called **postdiction** and is used to isolate those factors which are felt to be important in shaping development. In the case of the post-diction of the development of, for example, an urban area from 1950 to 1970, the 1970 situation may be used for predicting future development so that appropriate planning action may be taken.

There has already been some discussion of man's game against his environment, and it is not surprising that simulated game situations, with some emphasis on the operations of chance factors, have been used to obtain a better understanding of the workings of reality. A good example is to be found in Rex Walford's *Railway Pioneers* (Longman, 1972), which illustrates a large number of factors that affect route location. These include the search for an elusive 'least cost' route through mountains and plateaux, the promise of high revenues by reaching Californian goals, and the chance interventions of storm, espionage, sabotage, Indian attack, and failure of supplies. In competition against these odds and against each other, five or six railroad companies face these simulated hazards in pursuit of profit and prestige. An interesting point is that many of the problems that arise are not disclosed to the competing companies until the event occurs, and even then they are randomly selective in their effect. These hazards are revealed by the turn of a 'chance card' and indicate the importance attached to such random and unexpected occurrences in the development of the railroad pattern.

Descriptive models of behaviour, such as the farming one given in Fig. 4.6, build in the possibility of chance variation by emphasizing the filtering effect of the farmer's information field. All farmers have only imperfect knowledge of all the factors that affect their decisions. Each perceives his complete environment in an individual way, and bases his decisions on the often very imperfect information that his perception has provided. Thus, where a large number of farmers are operating in an area of little variation in relief, climate, and soils, there will tend to be quite a wide spectrum of decisions made. Most decisions will tend to group around a

Fig. 4.6 *A descriptive model of behaviour: a farmer's land use decision-making*

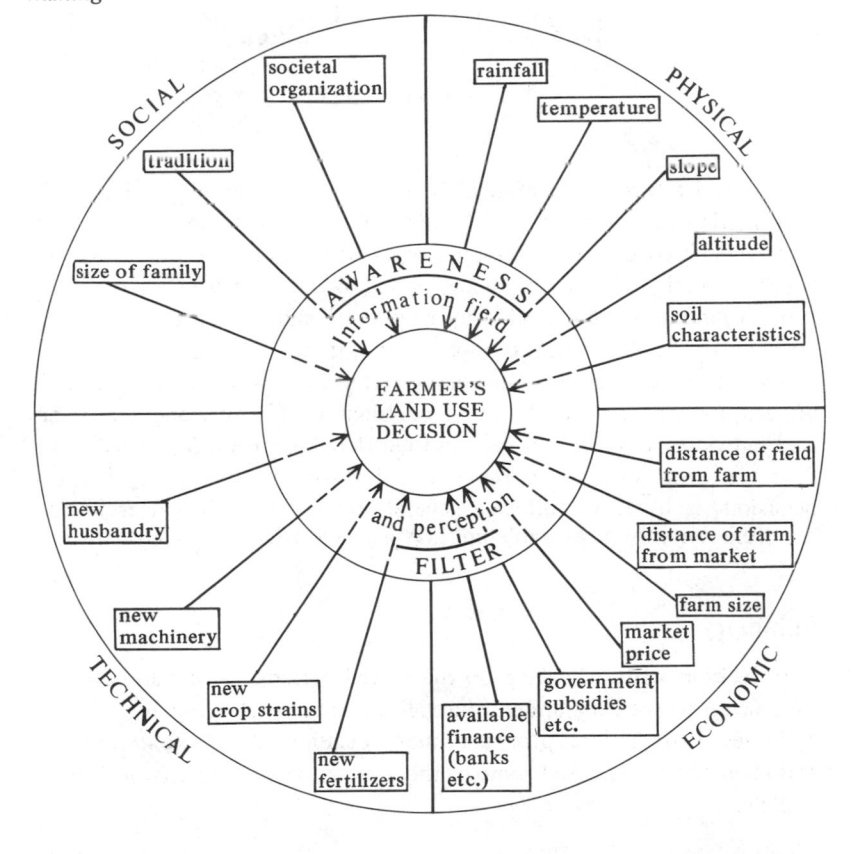

Fig. 4.7 *A fictitious example of normally-distributed decision-making where the decision involves choice of certain combinations of grass and cereals*

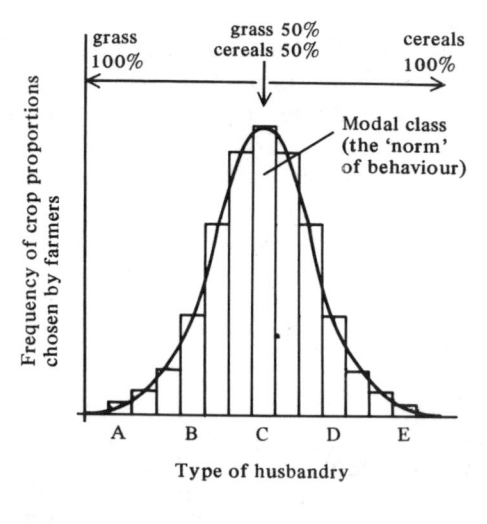

'norm', and a few vary more widely. Fig. 4.7 shows this diagrammatically. The distribution shown here is statistically a normal one, but it may be that where two or more different types of husbandry are present, there will be several peaks of frequency. In Fig. 4.7 we can see that the mixed type of husbandry represented by column 'C' is the one favoured by most farmers. At the extremes it is apparent that farmers choosing entirely grass or entirely cereals are few in number. Thus even with this simple model we can see that a wide choice of decisions is available, and that in any area individual farmers will vary in their decisions, though only a few will depart a great deal from the norm. This means that decision-making behaviour is likely to follow a general and recognizable pattern, and is therefore not only measurable but may also allow itself to be modelled.

Summary

Any decision with a bearing on the spatial patterns of human activity is of concern to the geographer. Since all that is studied in human geography is the result of the long process of man's decisions about his environment, this chapter has discussed some of the more important aspects of decision-making.

These decisions have to be taken in any case—whether rashly or sensibly —by individuals or by groups. By studying how decisions derive from the perception of the environment and of distance, geographers will be excellently placed to participate in planning for society's future.

In reading this book you will have realized why, if rigorous and significant analysis and hypothesis-testing are to be carried out, mathematical and scientific techniques must be applied to geographical data. The other books in this series explain in detail what these methods are and how they may be used.

Exercises

1. (a) List all the Greater London Boroughs alphabetically on a single sheet. Present this to a group with no prior knowledge of the exercise, and ask each person to list the top ten and bottom ten in order of their preference for each area as a place of residence. (If the group is not familiar with London, you may decide to use counties or some other familiar areas.) Add together all the choices given.

   (b) Maps may then be drawn of Greater London to show:
   (i) residential desirability;
   (ii) residential undesirability.

   (c) Such maps may then be related to existing maps of rateable value, social standing, and so on.

   (d) Some attempt may be made to see whether the pattern that results has a tendency to be annular (Burgess rings), sectoral (Hoyt model), or multi-nuclear (Harris and Ullmann model) in form. For a more detailed description of these models see Garner in Chorley and Haggett (1968), or Everson and FitzGerald (1972).

   (e) Co-operation with a group elsewhere in London could provide data for an interesting comparison.

2. Examine the family household movements over the past 20 years or so of a group of people personally known to you. Calculate the mean distance travelled. Suggest how this mean distance might vary with variation in income or social status.

References

1. Brookfield, H.C., 'On the Environment as Perceived', in Board, C., *et al.* (eds.), *Progress in Geography,* Vol. 1 (Arnold, 1969), p.63.

2. Brookfield, H.C., *op cit.*, pp.64—5.
3. Saarinen, T.F., *Perception of Drought Hazard on the Great Plains,* Chicago Research Paper No.106 (University of Chicago, 1966), page 76. Chapter 4 is also reprinted in Ambrose, P. (ed), *Analytical Human Geography* (Longman, 1969), pp.180—96.
4. Abler, R., Adams, J.S., and Gould, P., *Spatial Organisation* (Prentice-Hall, 1971), p.518.
5. FitzGerald, B.P. unpublished results of a survey of the families of geography students at St. Dunstan's College, Catford (1969).
6. White, Rodney, 'Mental Maps of Britain', *New Society*, 327 (2 Jan., 1969).

Further reading for Chapter 4

Abler, R., Adams, J.S. and Gould, P., *Spatial Organization* (Prentice-Hall, 1971), especially Chapters 11, 12, and 13.

Ambrose, P. (ed.), *Analytical Human Geography* (Longman, 1969), especially Sections 5 and 6.

Board, C., *et al.* (eds.), *Progress in Geography* (Arnold, 1969—72), various articles in each volume.

Everson, J.A., and FitzGerald, B.P., *Inside the City* (Longman, 1972), especially Chapters 1, 2, 11, 12, and 13.

Haggett, P., *Geography: a Modern Synthesis* (Harper and Row, 1972), especially Chapters 9 and 15.

Haggett, P., *Locational Analysis in Human Geography* (Arnold, 1965), especially pp.56—60, 79—86.

Lowenthal, D. (ed.), *Environmental Perception and Behaviour*, Chicago Research Paper No.109 (University of Chicago, 1967).

Saarinen, T.F., *Perception of the Drought Hazard on the Great Plains,* Chicago Research Paper No.106 (University of Chicago, 1966).

Berry, B.J.L., *et al.* (eds.), *Spatial Analysis* (Prentice-Hall, 1968), especially Hägerstrand, Torsten, 'A Monte Carlo Approach to Diffusion', pp.368—84, (reprinted from *European Journal of Sociology,* 6 (1965).

Bunge, W., *Theoretical Geography*, Lund Studies in Geography, (Gleerup, 1966), Chapters 2 and 7.

Everson, J.A., and FitzGerald, B.P., *Settlement Patterns* (Longman, 1969), Chapter 11.

Haggett, P., *Locational Analysis in Human Geography* (Arnold, 1965), pp.38, 53—5.

Bibliography

* Abler, R., Adams, J.S., and Gould, P., *Spatial Organization* (Prentice-Hall, 1971).

* Ambrose, P., (ed.), *Analytical Human Geography* (Longman, 1969).

‡ Berry, B.J.L., *et al.* (eds.), *Spatial Analysis* (Prentice-Hall, 1968)

+ Board , C., *et al.* (eds.), *Progress in Geography*, Vols. 1–4 (Arnold, 1969–1972).

+ Bunge, W., *Theoretical Geography*, Lund Studies in Geography (Gleerup, 1966).

+ Chorley, R.J., and Haggett, P. (eds.), *Models in Geography* (Methuen 1967).

* Chorley, R.J., and Haggett, P. (eds.), *Socio-Economic Models in Geography*, reprint of parts I and III of *Models in Geography* (1967), (Methuen, 1968).

+ Cole, J.P., and King, C.A.M., *Quantitative Geography* (Wiley, 1968).

✻ Everson, J.A., and FitzGerald, B.P., *Inside the City* (Longman, 1972).

* Everson J.A., and FitzGerald, B.P., *Settlement Patterns* (Longman, 1969).

* Haggett, P., *Geography: a Modern Synthesis* (Harper and Row, 1972).

+ Haggett, P., *Locational Analysis in Human Geography* (Arnold, 1965).

‡ Harvey, D., *Explanation in Geography* (Arnold, 1969).

+ Lowenthal, D., (ed.), *Environmental Perception and Behaviour*, Chicago Research Paper No.109 (University of Chicago, 1967).

* Morrill, R., *The Spatial Organisation of Society* (Wadsworth, 1970).

* Open University, *Understanding Society* (Macmillan, 1970).

* Theakstone, W.H., and Harrison, C., *The Analysis of Geographical Data* (Heinemann, 1970).

* Tidswell, W.V., and Barker, S.M., *Quantitative Methods: an Approach to Socio-economic Geography* (University Tutorial Press, 1971).

* Toyne, P., and Newby, P.T., *Techniques in Human Geography* (Macmillan, 1971).

‡ Yeates, M.H., *An Introduction to Quantitative Analysis in Economic Geography* (McGraw-Hill, 1968).

Note:
* Good clear outlines of subject under discussion, not too advanced for sixth-form and college students, although parts may be too demanding
+ Rather more advanced texts, but still worth referring to and containing sections that may be read easily.
‡ Advanced texts from which careful selection may be made.

Contents of Science in Geography, books 2, 3, and 4

S.I.G. 2   Data collection

Richard Daugherty

S.I.G. 3   Data description and presentation

Peter Davis

S.I.G. 4   Data use and interpretation

Patrick McCullagh